The Government of Science in Britain

Readings in Politics and Society

GENERAL EDITOR: Bernard Crick

Head of the Department of Politics and Sociology, Birkbeck College, University of London

Already Published

W. L. Guttsman, *The English Ruling Class*

A. J. Beattie, *English Party Politics* (two volumes)

Frank Bealey, *The Social and Political Thought of the British Labour Party*

Krishan Kumar, *Revolution*

Edmund Ions, *Modern American Social and Political Thought*

W. Thornhill, *The Growth and Reform of English Local Government*

Forthcoming titles

N. D. Deakin, *Race in British Politics*

Maurice Bruce, *The Rise of the Welfare State*

Previously published in this series by Routledge & Kegan Paul

David Nicolls. *Church and State in Britain Since 1821*

The Government of Science in Britain

Edited by
J. B. Poole and Kay Andrews
House of Commons Library

Weidenfeld and Nicolson
5 *Winsley Street London W1*

ISBN 0 297 99501 4

Printed in Great Britain by
Cox & Wyman Ltd
London, Fakenham and Reading

Contents

Contents

General Editor's Introduction

The purpose of this series is to introduce students of society to a number of important problems through the study of sources and contemporary documents. It should be part of every student's education to have some contact with the materials from which the judgements of authors of secondary works are reached, or the grounds of social action determined. Students may actually find this more interesting than relying exclusively on the pre-digested diet of textbooks. The readings will be drawn from as great a variety of documents as is possible within each book: Royal Commission reports, Parliamentary debates, letters to the Press, newspaper editorials, letters and diaries both published and unpublished, sermons and literary sources, etc., will all be drawn upon. For the aim is both to introduce the student to carefully selected extracts from the principal books and documents on a subject (the things he always hears about but never reads), and to show him the great range of subsidiary and secondary source materials available (the memorials of actors in the actual events).

The prejudice of this series is that the social sciences need to be taught and developed in an historical context. Those of us who wish to be relevant and topical (and this is no bad wish) sometimes need reminding that the most usual explanation of why a thing is as it is, is that things happened in the past to make it so. These things might not have happened. They might have happened differently. And nothing in the present is, strictly speaking, *determined* from the past; but everything is limited by what went before. Every present problem, whether of understanding or of action, will always have a variety of relevant antecedent factors, all of which must be understood before it is sensible to commit ourselves to an explanatory theory or to some course of practical action. No present problem is completely novel and there is never any single cause for it, but always a variety of conditioning factors, arising through time, whose relative importance is a matter of critical judgement as well as of objective knowledge.

The aim of this series is, then, to give the student the opportunity to examine with care an avowedly selective body of source materials. The topics have been chosen both because they are of contemporary importance and because they cut across established pedagogic boundaries between the various disciplines and between courses of professional instruction. We hope that these books will supplement, not replace, other forms of introductory reading; so both the length and the character of the Introductions will vary according to whether the particular editor has already written on the subject or not. Some Introductions will summarise what is already to be found elsewhere at greater length, but some will be original contributions to knowledge or even, on occasions, reasoned advocacies. Above all, however, I hope that this series will help to develop a method of introductory teaching that can show how and from where we come to reach judgements that are to be found in secondary accounts and textbooks.

Bernard Crick

Acknowledgements

The authors gratefully acknowledge the permissions to quote extracts in the book granted by the following individuals, journals, and institutions: the Controller, Her Majesty's Stationery Office; the Science Research Council; the Organisation for Economic Cooperation and Development; the Editor-in-Chief of *Impact of Science on Society* for an article in vol. III, no. 4, 1952, by M. Philips Price (Unesco); the Conservative Political Centre; the Labour Party; the Confederation of British Industry (for material published by the former Federation of British Industries); the Editor of *Nature*; the Editor of *Electronics & Power*; Birkbeck College, University of London; the Association of Scientific, Technical and Managerial Staffs (for material published by the former Association of Scientific Workers); Harrison & Sons Ltd.; the Rt. Hon. Lord Hailsham of Marylebone, the University of Southampton and Esso Petroleum Co. Ltd.; Professor J. P. M. Tizard; the British Association for the Advancement of Science; Roger Williams, Esq. and the Royal Institute of Public Administration; the Rt. Hon. R. H. S. Crossman, M.P. and *Encounter*; Austen Albu, Esq., M.P. and *Minerva*; S. A. Walkland, Esq. and the Editor of *Parliamentary Affairs*; The Economist Newspaper Ltd.; the Lord Blackett, O.M., C.H., F.R.S.; The Observer Ltd.; *New Statesman*; Routledge & Kegan Paul Ltd.; *The Spectator*; A. D. Peters & Company (for the extract from Sir Julian Huxley's 'Scientific Research and Social Needs'); and *New Scientist* (for an extract from Nigel Calder's 'Parliament and Science').

The authors especially thank Dr Roy MacLeod for his help and advice. They are also glad to acknowledge the help and encouragement they received from their colleagues in the House of Commons Library: David Holland, the Librarian; David Menhennet,

ACKNOWLEDGEMENTS

Deputy Librarian (who helped in particular with the chapter on
the Scientific Civil Service); Reg Brown, Senior Library Executive; Jane Brewster; and Eric Pound.

List of Documentary Sources

Part One: Historical Background

1 Central Organization: 1875–1919

1.1 **SOME PROPOSALS FOR REFORM, 1875–1903**

1.1.1 Eighth Report of the Royal Commission on Scientific Instruction and the Advancement of Science, under the chairmanship of the seventh Duke of Devonshire, C. 1298, 1875, pp. 45–6.

1.1.2 Michael Foster, 'The State and Scientific Research', *Nineteenth Century*, May 1904, vol. 55, pp. 742–5.

1.1.3 Sir Norman Lockyer, Presidential Address to the British Association, 1903, 'The Influence of Brain Power on History', British Association Annual Report for 1903, pp. 9, 26–8.

1.2 **STEPS TOWARDS THE STATE SUPPORT OF RESEARCH**

1.2.1 Sir Richard Glazebrook, FRS, a lecture delivered at the National Physical Laboratory on 23 March 1933.

1.2.2 First Annual Report of the Medical Research Committee for 1914–15, Cd. 8101, 1914–16, pp. 5–8.

1.3 **SCIENCE IN WARTIME: THE CREATION OF THE DEPARTMENT OF SCIENTIFIC AND INDUSTRIAL RESEARCH**

1.3.1 Board of Education, Proposals for a National Scheme of Advanced Instruction and Research in Science, Technology and Commerce, April 1915, Public Record Office Ed. 24/1581.

1.3.2 Deputation from the Royal Society and the Chemical Society to meet the Presidents of the Board of Education and the Board of Trade, 9 May 1915, Board of Education, Public Record Office Ed. 24/1579.

1.3.3 Sir William Ramsay, 'Science and the State', *Nature*, 20 May 1915, vol. 95, pp. 309–11.

1.3.4 H. G. Wells and Professor J. A. Fleming, letters to *The Times*, 11 and 15 June 1915.

2 Central Organisation: The Interwar Years

6 The Scientific Civil Service

7 Parliament and Science

7.2.3   M. Philips Price, 'The Parliamentary and Scientific Committee of Great Britain', *Impact of Science on Society*, Winter 1952, vol. 3, no. 4, pp. 269–71.

7.2.4   S. A. Walkland, 'Science and Parliament: the Origins and Influence of the Parliamentary and Scientific Committee II', *Parliamentary Affairs*, 1963–4, vol. 17, pp. 399–400.

7.3   SELECT COMMITTEES

7.3.1   N. J. Vig and S. A. Walkland, 'Science Policy, Science Administration and Parliamentary Reform', *Parliamentary Affairs*, 1965–6, vol. 19, pp. 289–90.

7.3.2   Austen Albu, MP, 'The Member of Parliament, the Executive and Scientific Policy', *Minerva*, Autumn 1963, vol. 2, no. 1, pp. 17–20.

7.3.3   Conservative Political Centre, *Change or Decay : Parliament and Government in Our Industrial Society*, January 1963, pp. 11–12.

7.3.4   Study of Parliament Group, Memorandum Submitted to the Select Committee on Procedure, Appendix 2 to the Fourth Report from the Select Committee on Procedure, House of Commons Paper 303, 1964–5.

7.3.5   Parliamentary and Scientific Committee, Subcommittee on Parliament and Science, Memorandum Submitted to the Select Committee on Procedure, Appendix 3 to the Fourth Report from the Select Committee on Procedure, House of Commons Paper 303, 1964–5.

7.3.6   Nigel Calder, 'Parliament and Science', *New Scientist*, 28 May 1964, vol. 22, p. 535.

7.3.7   House of Commons Debate, 14 December 1966. Speech by Rt. Hon. R. H. S. Crossman, MP, Hansard, vol. 738, col. 479–86.

7.3.8   'Chink in the Flood Gates', *Nature*, 24 December 1966, vol. 212, p. 1395.

7.3.9   Roger Williams, 'The Select Committee on Science and Technology: the First Round', *Public Administration*, Autumn 1968, vol. 46, pp. 309–11.

7.3.10 House of Commons Debate, 29 October 1969. Speech by Mr Eric Lubbock, MP, Hansard, vol. 788, col. 1012–6.

Introduction

SCOPE AND ORGANISATION

Our object in this book has been to illustrate how the government of science in Britain has developed over the past half century. Alternatively, we could express our aim by saying that we have tried to show how British science policy has emerged over this period. It is arguable that our application of so relatively recent a coinage as the term 'science policy' to the early part of this century is unwarranted. However, perhaps by very virtue of its newness, it has a commendable elasticity, and sufficiently so to cover all that we would consider relevant to science and government. There cannot yet be said to be universal agreement on what science policy is but the definition put forward by the Organisation for Economic Co-operation and Development is a convenient one that we shall adopt:[1]

By a national science policy is meant the deliberate attempt of a government to finance, encourage and deploy the scientific resources of the country – trained research workers, laboratories, equipment – in the best interests of national welfare. Such a concept of policy presupposes a recognition of science as a powerful influence on some or many aspects of national life – cultural, social, health, defence, economic, etc. – which until recently could not generally be assumed.

No collection of documents could do justice to the profound changes in the relations between science and government during the twentieth century. With this in mind, we have not attempted to be comprehensive, in that we have not tried to isolate and follow through the development of every aspect of the subject; rather we have selected a few of the outstanding features and emphasised

[1] *Fundamental Research and the Policies of Governments* (Paris, 1966), p. 16.

I

these. One of the major drawbacks we have encountered in making this selection is the absence of any authoritative, historical account of the organisation of British science over the past century. We do not offer this book as any substitute for such a study.[2] We would not even suggest that this book is necessarily suitable for cover-to-cover reading; we feel that it might be most appropriately used as a source book, where readers may look to find extracts, placed in context, from those documents to which the texts refer but rarely quote.

We have been limited in our choice of documents by two factors. First, considerations of space; to cover the subject adequately, to discuss in detail the evolution of policy-making would require a multi-volume work. In the event, the choice of documents was resolved not so much by what should be included, as by what must not, on any account, be left out. Secondly, we have been limited by what was available. The skeleton of the story can be found in Parliamentary material; departmental material is often more elusive. As the study is concerned with the response of government to science and technology it is proper that government should speak for itself. The public reaction of the scientific community to the actions of governments is transmitted by journal editorials, letters to the press, books, and, to a degree, through Parliament, and these are all drawn upon appropriately.

Harder still to track down and perhaps more intimate are the comments, discussions and suggestions of the scientists and administrators, whose persuasive enthusiasm and influence can often be discerned at the policy-making level. Men such as Lord Balfour are rare at any time; in his papers we can see the dedication of a true 'statesman of science'. Unfortunately there are few such accessible collections of papers. The world of the private scientist and of the civil servant remain, for the most part, closed, save for a few tantalising glimpses. The task of assessing the influence of the scientist in the political arena remains to be done.

The period we have covered ranges, broadly, from the first world war to the present day. The two world wars that occurred during this period constituted turning points in the organisation, mobilisation and exploitation of science in Britain. From the first emerged

[2] An exception must be made for N. J. Vig's *Science and Technology in British Politics* (Pergamon Press, Oxford, 1968), although it covers in detail only a relatively short period: the early 1960s.

the germ of national science planning. The second gave tangible form to the governmental interventionism which has subsequently come to play such a noticeable part in the science policy of Britain and other countries. The terminal date in our selection of extracts is mid-1970.

Apart from the convenience of the deadline it is also an aptly appropriate date. The 1970 general election brought to an end the second postwar Labour government and returned a new Conservative administration, having as one of its electoral pledges the aim of reforming and reducing the machinery of government which owed much to the innovations introduced by Labour after the general elections of 1964 and 1966. Yet whatever changes are introduced eventually by the new Conservative government,[3] insofar as they impinge on science policy, they are likely to echo to a degree the (now abortive) plans of the defeated Labour administration to restructure its biggest creation, the Ministry of Technology. Thus 1970 is likely to be seen in retrospect as marking the end of a major phase in postwar British science policy.

For the most part we have restricted ourselves to *civil* science, as distinct from that concerned with defence; there are two main reasons for this: organisationally it has been relatively easy to distinguish between the two in Britain, and secondly there is much less usable material available relating to defence matters than there is to civil science.[4] The place of science in war is essentially a separate phenomenon which, within the limits of this book, we would not have been able to do justice to.[5] Thus we have chosen to refer briefly to those research developments that proved critical in the winning of the two world wars and in determining the priorities of peacetime. We have further excluded any detailed consideration of atomic energy, whether for military or peaceful purposes, since this does not lack its own admirable, definitive – although still to be

[3] The first changes to be announced were given in the White Paper *The Reorganisation of Central Government*, Cmnd. 4506, October 1970.
[4] The report of the Select Committee on Science and Technology on Defence Research (House of Commons Paper 213 of 1968–9) is a notable addition to the literature.
[5] See, for example: Earl of Birkenhead, *The Prof. in Two Worlds : the Official Life of Professor F. A. Lindemann, Viscount Cherwell* (Collins, London, 1961); R. W. Clark, *The Rise of the Boffins* (Phoenix House, London, 1962); M. M. Postan, D. Hay and J. D. Scott, *History of the Second World War. Design and Development of Weapons, Studies in Government and Industrial Organisation* (H.M.S.O., London, 1964).

3

completed – history.[6] We have endeavoured to cover the post-second world war development of manpower policy, but we have not dealt with scientific and technological education and training in detail.[7]

While we try to forestall criticisms of over-emphasis, we have found it possible to achieve a greater continuity by concentrating in our selection on the organisation for and the performance of the physical sciences and technology. We have also tried, however, to make appropriate reference to analogous developments in other areas, for instance the medical and the agricultural sciences.

We are all too conscious that the international aspects of science and technology have had to be excluded. This is yet another area lacking a thorough study. Nevertheless, a further justification for the investigation in depth of the institutionalisation of science in Britain is that the Research Council formula was widely adopted throughout the Empire (Australia, India and South Africa, for instance).

The book is divided into two parts. The first deals chronologically with the organisation of science from 1875 on. The second part is thematic in arrangement; here we isolate and concentrate upon four 'elements' that have been of particular importance in the evolution of science policy in this country: research and development, manpower resources, the Scientific Civil Service, and the role of Parliament. By 'science' we are to be interpreted as including also 'technology' (a very lopsided picture would have resulted by attempting to separate them systematically), and insofar as the latter extends to, for instance, government policy for the promotion of industrial research and development, we have considered this to be relevant to the book.

We have made it clear that we do not propose to tell the detailed history of the manner in which the government of science has developed. We propose, however, in the next few pages, to outline some of the major developments and administrative changes dating from the nineteenth and early twentieth century on which the present organisation of science policy rests. This background,

[6] M. Gowing, *Britain and Atomic Energy, 1939–1945* (Macmillan, London, 1964).
[7] Representative studies are: M. Argles, *South Kensington to Robbins* (Longmans, London, 1964); D. S. L. Cardwell, *The Organisation of Science in England* (Heinemann, London, 1957); G. L. Payne, *Britain's Scientific and Technological Manpower* (Stanford University Press, Stanford, 1960).

particularly emphasising the earlier developments, will enable readers to place the quoted extracts in some context. The second part of the introduction explains briefly why we have isolated the four characteristics described above. Additionally each chapter has a prefatory note.

THE HISTORICAL BACKGROUND

Science and government in the nineteenth century

The growth and institutionalisation of scientific research has its roots in the eighteenth and the early nineteenth century. What now appears to us as a highly expensive and sophisticated phenomenon, began as a modest, individual and haphazard activity. The first indications of the government's awareness of the immediate useful-ness of science to the community began with such initiatives as the founding of the Royal Observatory in Greenwich in 1675. In the first half of the nineteenth century, however, there was a significant change in the support of science. At the beginning of the century there was a feeling that, compared with the continent, and parti-cularly France and Germany, science was 'in decline' in England.[8] Whether this was in fact so, in terms of scientific productivity, or in terms of popular interest, is not entirely clear, but the effect of this damaging and demoralising comparison provided the stimulus for the creation of the British Association for the Advancement of Science in 1831 and a climate for reform in the Royal Society in the 1840s. The motives of those who began the movement for re-appraisal and reform, such as Charles Babbage and David Brewster, and the truth of these allegations about the state of science in England, still provide a fruitful field for speculation. One signifi-cant by-product of the self-examination of the scientists came in 1850, with the government's decision to place an annual grant of £1,000 with the Royal Society, to encourage the pursuit of research. The annual grant was later increased to £4,000. Between 1850 and 1914 it assisted nearly a thousand scientific men in their re-search with grants ranging from £50 to £300.[9] The significance of

[8] Charles Babbage, *Reflections on the Decline of Science in England* (Fellowes, London, 1830).
[9] R. M. MacLeod, 'The Royal Society and the Government Grant: Notes on the Administration of Scientific Research, 1849–1914', *Historical Journal*, 1971, vol. 14, no. 2, pp. 323–58.

this award, which was largely the result of the enthusiasm of Sir Robert Peel, was that, together with the smaller amounts available from the British Association, it remained the only substantial public source of money for independent research (unlike schemes solely designed to reflect excellence, such as the medal system of the Royal Society). It marked, however, a significant departure from the traditional criterion of 'utility' which had influenced the setting up of the Geological Survey in 1832, the Museum of Economic Geology in 1841, and the laboratory dealing with the adulteration of imported perishable goods attached to the Inland Revenue Department in 1843. In time, these first tentative steps towards state-run scientific investigations of economic importance assumed a permanent and expensive character. The Inland Revenue Laboratory later became the Laboratory of the Government Chemist, which in 1911 became a department of the Treasury to test agricultural products, fertilisers, and to investigate water and atmospheric pollution.

While the award of the government grant to the Royal Society was a marked diversion from the norm of utilitarianism, it did not presume a change in the government's attitude. However, the phenomenal success of the Great Exhibition of Industry of All Nations of 1851 marked a turning point in public attitudes towards science and technology. The Exhibition, largely the result of the combined enthusiasm and administrative genius of Prince Albert, Sir Robert Peel, Lyon Playfair and Henry Cole, proved a convincing demonstration of British technical and engineering skill.[10] Playfair seized the opportunity to make the Exhibition not a passing monument but a permanent endowment for scientific education through the creation of new institutions. The only scientific colleges in existence at this time were the Royal College of Chemistry in Oxford Street, established in 1845, and the Government School of Mines and Science Applied to the Arts in Jermyn Street, which in 1851 became the School of Mines. There was a surplus of £186,000 from the Great Exhibition and this money, plus another £150,000 voted by Parliament, was used to purchase the South Kensington Estate. The two science colleges were transferred to South Kensington and by 1872 formed the basis of the Royal College of Science – now the Imperial College of Science and

[10] See, for example, Christopher Hobhouse, *1851 and the Crystal Palace* (John Murray, London, revised edn. 1950).

Technology. The other indirect results of the Exhibition later in the century were the building of the Natural History Museum and the creation of the Imperial Institute, designed to promote knowledge of the raw materials of the Empire and of their uses.

The sense of public excitement generated by the 1851 Exhibition and enshrined in the science schools was complemented by substantial changes in the organisation and administration of education. In 1853 the Science and Art Department had been formed, with Lyon Playfair as science secretary. However, according to one view, the Department began to make an impact only after the introduction of the 'payment by results' scheme, whereby trained science teachers were paid according to the examinations success of their pupils. Although a crude scheme in design, it was successful in that the number of pupils examined in science increased from about 1,300 in 1861 to 100,000 in 1887. At the same time there were reforms in the older universities, which, following the Royal Commissions of 1852 had revised the examination requirements. In Oxford, honours schools were created in theology, law, history and natural sciences, and analogous innovations were introduced at Cambridge in natural and moral sciences.

British complacency following the evidence of the Great Exhibition was, however, shattered by the revelations of the Paris International Exhibition of 1867. Playfair, on return from the Exhibition, emphasised that international opinion believed that Britain had made little progress in recent years. The public outcry following Playfair's accusations resulted first in the decision to create a Select Committee on Scientific Instruction in 1868, which recommended among other things the foundation of new liberal colleges along the lines of Owens College, Manchester (founded in 1851). The issues raised by the 1868 Committee were raised again in 1871 with the decision (under pressure from a Committee of the British Association, led by Colonel Alexander Strange) to set up a further commission devoted to 'scientific instruction and the advancement of science'. The Devonshire Commission (after the chairman, the seventh Duke of Devonshire) provided a unique opportunity for the thorough review of scientific organisation, research and teaching. Among other things the Commission recommended the radical reform of secondary education. Most importantly, however, the Commission agreed with Strange's proposals for state-run laboratories, for increased research grants for private scientists,

7

and a Ministry of Science and Education assisted by a Council of Science.

Despite the dedication of Norman Lockyer, secretary to the Commission, and the optimism of those scientists and statesmen who gave evidence, the Royal Commission in fact achieved little. The opportunity for the state to endow a scientific university was missed, and the concept of a Ministry of Science had to wait fifty years before its partial implementation in the Department of Scientific and Industrial Research.

The Devonshire Commission brought to a head the contemporary debate on the position of the scientist and the value of scientific research to the community. With the founding of the journal *Nature* in 1869[11] the scientific community, under the energetic leadership of Lockyer, had a platform for its concern. The 'movement for the Endowment of Research',[12] which aimed at making research scientists independent of teaching by the grant of research fellowships, found a welcome in the columns of *Nature*. However, neither the Devonshire Commission nor the efforts of *Nature* and of the movement for the Endowment of Research were able to improve the economic or social position of the scientist. Despite the growth of opportunities in the laboratories of the new provincial universities, the research worker remained an isolated creature, and the economic difficulties of pursuing his work remained unchanged.

As far as it is possible to generalise on what is still a virtually unexplored field, it would seem that there were three elements on which nineteenth-century government 'science policy' was based. First, awards of money were made in support of pure research, either as grants to individuals or as grants administered by the Royal Society for the funding of expeditions or specific pieces of research, such as solar physics. There were also international expeditions (for instance, the *Challenger* expedition which explored the Pacific Ocean and crossed the Antarctic Circle between 1872 and 1876 – its report filled 50 volumes), which consumed a perhaps disproportionate amount of money and were in many ways analogous to the present-day international space race. Treasury

[11] See the commemorative articles by R. M. MacLeod in the centenary issue of *Nature*, 1 November 1969, vol. 224.
[12] C. Appleton, *et al.*, *Essays on the Endowment of Research* (King, London, 1876); R. M. MacLeod, *Minerva*, 1971, vol. 9, no. 2, pp. 197–230.

relations with the scientific societies in reviewing grants for such expeditions were affected by the natural reluctance of civil servants to invest in such an unaccountable phenomenon as 'scientific research', and by the dangerous liaison (under first the Science and Art Department and then the Board of Education) of research and education, since the massive costs of education were a constant tribulation to the Treasury. On the other hand, there was an explicit faith in the integrity of the Royal Society. This alliance with the establishment, and the fear of being identified as politically partisan, prevented the Royal Society from more energetically supporting the community's demands for more money. At the same time, money was available for expensive utilitarian research, such as the Geological Survey, the Ordnance Survey, and the technical branches (such as in public health) of the Local Government Board, the Home Office (such as explosives) and the Board of Trade (such as lighthouses). In each of these departments the 'scientific expert' was becoming a common phenomenon. The work of such pioneers as Sir John Simon, the first medical officer of the Privy Council, who was responsible for scientific investigations into disease and public health,[13] and Dr Angus Smith, the government's first Alkali Inspector, marked the beginning of a new era in public service. The public health inspectors, factory inspectors, scientific advisers to the Board of Trade and expert advisers to the Royal Commissions formed a first impressive section of the new expert class.

In 1899, however, with the creation of the National Physical Laboratory, the state came into a new relationship with science. The 1890s had seen increasing anxiety about German industrial competition. In 1895 the Physikalische-Technische Reichsanstalt was opened at Charlottenburg and was soon an essential part of the German scientific and industrial complex.[14] While *The Times* urged the increased application of science to industry, *Nature* criticised the attitude of the 'inexpert' government. In 1896, E. E. Williams's *Made in Germany* appeared, and the fear of German 'efficiency' reached a new intensity: 'Blame was shared out between

[13] R. Lambert, *Sir John Simon* (MacGibbon & Kee, London, 1963).
[14] For an analysis of German support for science see F. Pfetsch, 'Scientific organisation and science policy in Imperial Germany, 1871–1914: the foundation of the Imperial Institute of Physics and Technology', *Minerva*, October 1970, vol. 8, pp. 557–80.

apathetic history, inadequate secondary schools and universities unfitted for research.'[15] The Royal Society came in for a share of the blame too for its policy of electing non-scientific fellows.

In the midst of the growing unease about German industrial and military power, a government committee led by Sir Oliver Lodge, Lord Rayleigh and Francis Galton persuaded the Treasury of the national need for a physical laboratory, analogous to the Physikalische-Technische Reischsanstalt. At its opening the Prince of Wales described the significance of the National Physical Laboratory:

The object of the scheme is, I understand, to bring scientific knowledge to bear practically upon our every day industrial and commercial life, to break down the barrier between theory and practice, to effect a union between science and commerce . . . Does it not show in a very practical way that the nation is beginning to recognise that if her commercial supremacy is to be maintained, greater facilities must be given for the further application of science to commerce and manufacture?[16]

The success of the Laboratory and the vociferous agitation of the newly formed British Science Guild[17] prepared the way for a further injection of state aid to research following Lloyd George's Liberal budget of 1909, which was an unprecedented attempt to finance social reform on a national level. There were two aspects of Liberal policy which affected science: first, the decision to set up a Development Fund for the development of the agricultural resources of the country; and, second, the emergence of state-financed medical research, as a consequence of the National Insurance Act of 1911. The Development Fund of £2½ million was to provide for the 'proper scientific development of afforestation, agriculture and fisheries'.[18] The fund was to be administered

---

[15] E. E. Williams, *Made in Germany* (Heinemann, London, 1896).
[16] First Report of the Committee of the Privy Council for Scientific and Industrial Research, for the year 1915–1916, Cd. 8336, 1916, p. 5.
[17] The British Science Guild was formed in 1905 by Norman Lockyer to be a pressure group for science and scientific method in the public arena. Lockyer had originally intended it to act as a ginger group within the British Association, but the reluctance of that body to take the initiative forced him to form an independent group. In the next thirty years it did invaluable work in promoting a public awareness of scientific issues in research, teaching and industry. In 1936 it was finally absorbed by the British Association.
[18] J. Russell, *A History of Agricultural Science in Great Britain, 1620–1954* (Allen & Unwin, London, 1966).

by eight commissioners – seven part-time and unpaid, and one full-time and salaried. The first line of action was to set up a system of sound scientific investigation accompanied by a corresponding development of agricultural education under the direction of Daniel Hall.[19] During 1910–11 the Development Fund began research support for Rothamsted (soil and plant nutrition), Cambridge (plant breeding and animal nutrition) and Long Ashton (fruit and cider). By August 1914 twelve institutes and two minor research centres had been set up in England and Wales: four for plant sciences, six for animal sciences, two for fruit growing, one for dairying and one for agricultural economics.

Apart from the work of the Local Government Board, medical research was not supported by the state until 1911 and the passing of the National Insurance Act, as already mentioned. State aid for medical research grew out of Lloyd George's enthusiasm for insurance, based on the unfavourable comparisons made between England and the continent, where compulsory social insurance for sickness, accidents and old age existed in many countries.[20] The Act aimed at bringing the whole wage-earning population within a scheme of compulsory contributory health and unemployment insurance. The payments guaranteed, among other things, medical attendance and drugs. One penny per insured person was to go to a fund to be primarily directed towards tuberculosis research. By 1914 the annual income of the fund amounted to £55,000, and it became clear that the Committee which administered it was committed to a wide and ambitious programme of fundamental medical research. With the outbreak of war in 1914, the movement for increased and consistent support of research was inevitably accelerated.

From the first to the second world war

In his Rede Lecture in 1919, Lord Moulton described the impact of science on the war in the following words:

But for the stupendous advances that Science has made in times within the memory of many here present no catastrophe at once so wide-spreading and so deep-reaching could have happened. In scale and intensity alike this War represents the results of the

[19] H. E. Dale, *Sir Daniel Hall: Pioneer in Scientific Agriculture* (John Murray, London, 1956).
[20] J. Brand, *Doctors and the State* (Johns Hopkins Press, Maryland, 1965).

totality of scientific progress – it is the realisation of all that which the accumulated powers with which Science has endowed mankind can effect when used for destruction.[21]

Moulton went on to catalogue those scientific developments which had characterised the war and which would, in the future, mark it out from all other wars: inventions such as the tank, the aeroplane and the submarine, which had revolutionised the scope and speed of war and made the concept of 'total war' an immediate reality; inventions which had been used defensively against such devices; the first steps towards sound-ranging techniques, wireless and magnetic mines; and above all poison gas. Moulton's description of science and war would have been supported by the majority of his listeners. The war had proved to be a turning point in the use and organisation of science. The scientist's role in the development of weaponry was accepted, often reluctantly, into the entrenched traditions of the services.[22] Yet, as the war progressed, and the introduction and testing of weapons and devices assumed increasing importance, the professional scientist took his place next to the serving officer. But the essential difference of approach between the officer and the scientist could not be disguised.[23] The war brought the scientist into greater national prominence than ever before. Victory was seen not merely as a moral triumph but also as a vindicating of British ingenuity and application.

One immediate result of the war was to justify the assertions of the long line of critics of British science and technology. The first months of the war revealed not only severe shortages of scientists and engineers but also the great dependence of British industry on German drugs, dyestuffs and optical goods which were essential to the waging of war. To combat public anxiety and to try to solve some of the more pressing shortages, the government proposed a Council for Scientific and Industrial Research which would not only try to deal with the immediate problems but also prepare a scientific blueprint for peacetime. This proposal originally came from the government's Board of Education whose responsibility covered only England and Wales. Thus in order that the proposal

[21] *Science and War* (Cambridge University Press, 1919), pp. 6–7.
[22] R. M. MacLeod and E. K. Andrews, 'Scientific advice in the war at sea, 1915–1917: the Board of Invention and Research', *Journal of Contemporary History*, 1971, vol. 6, no. 2, pp. 3–40.
[23] Clark, *The Rise of the Boffins.*

should be applicable to the whole kingdom the device was adopted, in 1915, of creating a Committee of the Privy Council for Scientific and Industrial Research; this Committee was to be advised by a small but eminent Advisory Council. One major innovation to come from this advisory group was a scheme for the encouragement of industrial research associations, the so-called 'Million Fund'.

It quickly became apparent that it would be desirable to constitute a separate department in order to further the Advisory Council's work. Thus the Department of Scientific and Industrial Research (DSIR), having its own Parliamentary vote and responsible to Parliament through the Lord President of the Council, was formed at the end of 1916.

DSIR had three prime areas of responsibility: it controlled its own laboratories, numbering fifteen at the time of its dissolution and concerned with, for instance, the building industry, road research, fire prevention and fish preservation; it administered the research association scheme (the number of these has fluctuated: there were about twenty established between the wars and there were forty-six in 1969); and lastly the Department awarded research grants to postgraduate students in order to improve the country's scientific manpower stock. By the end of the war, the relationship between science and the state had subtly but irrevocably changed.

During the interwar years the organisation of science was marked by four outstanding contextual characteristics, some deriving from the nature of the scientific institutions themselves and others from external factors which directly affected the rate and direction of scientific development.

The first and most important external factor was the uncertain economic climate of the 1920s and early 1930s, which had a profound effect on the planning and execution of research programmes, in government, the universities and industry. The DSIR turned its attention towards the solution of those problems which would most contribute towards prosperity in peacetime. By the end of the war the Department had closely linked its own research priorities with social and economic goals. Research programmes were begun to stimulate the more efficient use of fuel, to develop home-grown timber and reduce the traditional reliance on Imperial and foreign supplies, to provide cheaper and more effective building materials and bring the 'Homes for Heroes' nearer reality, and to produce and preserve more homegrown foodstuffs. In addition, the DSIR also assumed responsibility for the National Physical Laboratory and the Geological Survey. Furthermore, through its system of postgraduate research awards it was committed to the encouragement of research workers.[24]

In 1918 the Machinery of Government Committee of the Ministry of Reconstruction, under Lord Haldane, had concluded its review, and had confirmed the wisdom of the DSIR Research Council arrangement, which left the Department answerable only to the Lord President of the Privy Council. The Haldane Committee recommended that the same formula be applied to both medical and geological research. To the Medical Research Committee, in a restricted position under the National Insurance Commission, this recommendation was welcome. In 1920 the Medical Research Committee became the Medical Research Council[25] and joined the DSIR under the benign patronage of Lord Balfour, then Lord President. The 'Haldane Principle' of the separation of research from departmental political control has become enshrined as one of those tenets of British government which are often quoted while their implications are rarely understood. It is only recently that the validity and applicability of this principle has come into question.[26]

By 1919 it seemed that the organisation of research was being rationalised. The onslaught of economic restriction in 1920-1, therefore, came as a serious blow. The postwar boom did not last long. The war had revolutionised prewar trade patterns and Britain was the loser: rising unemployment and a startling rise in prices in the winter of 1920-1 forced the government into policies of severe retrenchment. Research was a comparatively new and vulnerable factor in the budget. In addition to reducing the universities' new annual grant from £1,500,000 by £300,000 (at a time when the universities were overflowing with demobilised

---

[24] H. Melville, *The Department of Scientific and Industrial Research* (Allen & Unwin, London, 1962); H. F. Heath and A. L. Hetherington, *Scientific and Industrial Research* (Faber & Faber, London, 1946).

[25] For an account of Sir Walter Morley Fletcher's work as First Secretary of the MRC, see the biography by his wife, M. Fletcher, *The Bright Countenance: a Personal Biography of Walter Morley Fletcher* (Hodder & Stoughton, London, 1957). See also J. G. Crowther, *Scientific Types* (Barrie & Rockliff/Cresset Press, London, 1968).

[26] See Lord Zuckerman's 2nd TLS Lecture, *The Times Literary Supplement*, 5 November 1971, pp. 1385-8.

students), the Treasury demanded a reduction of £100,000 in the
DSIR's estimates between 1920–1 and 1921–2. The DSIR complied,
at the expense of many of its embryonic research programmes. For
the next five years the DSIR did not recover from this crippling
blow, and it was not until 1926 that its budget began to approach
parity with that of 1920–1. The results of these cuts, although
carried out in a gentlemanly fashion with much reluctance by the
Treasury and good grace by the DSIR, were to have a profound
influence on the capability of the Department to respond to changing demands and conditions of peacetime. For most of the 1920s it
could barely manage to keep its committed research programmes
running, and the effect on the research staff, constantly under the
threat of further reductions, was demoralising and counterproduc-
tive. While the 1930s brought some relief, it is interesting to
speculate on what the Department might have achieved under
more favourable circumstances.

The second external characteristic which influenced scientific
organisation was the growth of the postwar Empire through the
acquisition of prewar German territories, in East Africa. One
of the immediate problems besetting the government was how best
to develop these new dependencies. The interwar years, therefore, saw a new interest in the scientific development of the
Empire. A Colonial Research Committee was set up in 1920 to coordinate work for it. The Committee of Civil Research, created in
1925, grew out of revelations that the newly acquired African
territories would be lost without the serious application of scientific
methods in agriculture and medical services. Throughout this
period general research programmes were set up for agriculture,
medicine and nutrition, and the problem of Imperial development
through science became a favourite topic of government inquiry
and action.

The third characteristic of interwar science was an internal
feature of the development of the postwar scientific community:
the change in the status of the scientist. At the immediate end of the
war there were signs of self-interest and awareness among research
scientists in the decision to create a National Union of Scientific
Workers (later, the Association of Scientific Workers), which
would not only promote the economic interests of its members,
particularly those in the Civil Service, but would also guard the
interests of science itself. The Union took its place among the ranks

of the white collar trade unions, such as the Institution of Professional Civil Servants, emerging in the 1920s. Although the membership was never large, fluctuated throughout the 1920s, and nearly expired completely at the beginning of the 1930s, there is no doubt that the existence of a corporate body of scientists and technicians raised the collective morale of the scientific community and won for the scientist an organised voice in the political arena.

Fourthly, and finally, one further development which was to have a profound influence on the pattern of scientific thinking was the growing concern among scientists with the organisation and application of their work. The 1930s opened with international violence and economic confusion. For many people in Britain, including some notable scientists, it seemed that the order of Stalinist Russia was preferable to the economic and political anarchy of western Europe. The comparison stretched to include science, and in 1931 the relationship between science and the state in Marxist terms was fully explored at the Second International Congress of the History of Science and Technology.[27] The debate profoundly influenced many left-wing scientists in Britain. That science could be a tool of national development, and not merely the recipient of state support, was an argument readily accepted in the harsh environment of the 1930s with professional unemployment almost equalling that in yet more vulnerable sectors of society. The arguments for planning in science, supported by such venerable members of the scientific (and political) community as Sir Daniel Hall (see above), and as enthusiastically by many younger men, were put forward as applicable to all branches of science from medicine to agriculture.[28] The arguments raged throughout the 1930s, but with the coming of war and the promise of even greater technological advances the argument was silenced.

From the second world war

Many of the post-second world war changes in civil science centred about the DSIR. The other Research Councils remained essentially unchanged, although a new one, the Nature Conservancy, was

[27] N. I. Bukharin, *et al.*, *Science at the Crossroads* (F. Cass, London, new edn, 1971).
[28] For a study of the political attitudes of certain leading scientists in this period, see P. G. Wersky, 'British scientists and "Outsider" Politics, 1931–1945', *Science Studies*, January 1971, vol. 1, no. 1, pp. 67–83.

created in 1949. The structure of the DSIR remained similarly unchanged until 1956 when, in order to improve the direction of the research establishments under the Department's control, it was decided to change the Advisory Council into an executive Council for Scientific and Industrial Research, of which the Lord President remained chairman.

The DSIR was dissolved by the second postwar Labour government in 1965, its headquarters staff and most of its research stations going to form the new Ministry of Technology and others going to a new Science Research Council. From earlier in the 1960s it had been felt that some change in central organisation was required and the preceding government had set up a Committee of enquiry into the Organisation of Civil Science (the Trend Committee); this Committee had also recommended the disbanding of DSIR to form a Science Research Council, subsequently adopted as we have noted, and an Industrial Research and Development Authority, a concept which reemerged in a new proposal in 1970.

During the lifetime of the 1964–70 Labour government the Ministry of Technology was further expanded, most notably by the acquisition of the former Ministry of Aviation, the Ministry of Power together with a supervisory role over the Atomic Energy Authority, and the National Research Development Corporation – a creation of the first postwar Labour government intended to exploit inventions in the public interest through appropriate injections of public funds. Together with certain industrial sponsorship roles, e.g. for shipbuilding and the computer industry, the Ministry was, in effect if not in name, a Ministry for industry.[29] The Conservative government elected in 1970 honoured its pre-election pledge and turned much of the Ministry, plus the former Board of Trade, into a new Department of Trade and Industry.[30]

We have given precedence to the evolution of the DSIR and the post-Trend Report developments since these bulk large in the extracts that follow, but it is important not to assign DSIR an importance that it did not possess. Thus in 1963–4 the total government expenditure on civil science was £174·3 million, of which the major components were:

[29] M. Shanks, *The Innovators* (Penguin Books, Harmondsworth, 1967).
[30] See the articles in *The Sunday Times*, 27 February and 5 March 1972.

	£000
DSIR	25,143
Agricultural Research Council	7,189
Medical Research Council	7,033
Nature Conservancy	721
UK Atomic Energy Authority	45,000
National Institute for Research in Nuclear Science	7,800
Ministry for Aviation	37,778
Ministry of Agriculture, Fisheries & Food	3,139
Universities	31,000

It will be apparent that the Research Councils overall account for only a relatively small part of the total outlay. (For comparison, in the same financial year roads in England and Wales cost £151 million, the National Health Service £623 million and family allowances £141 million.)

With the passing of the Science and Technology Act 1965 the Research Councils were made answerable to the Secretary of State for Education and Science, instead of the Lord President of the Council, but were otherwise unchanged; the same Act created the Science Research Council, and the Natural Environment Research Council (absorbing the Nature Conservancy as a semi-autonomous entity). A Social Science Research Council was also formed in 1965.

An extract in chapter 4 below shows how the DSIR became involved with work on atomic weapons in the second world war. At the end of the war this responsibility was passed to the Ministry of Supply, a primarily defence-oriented department. In the course of time it appeared that atomic energy would have civil as well as military applications (electrical power generation in particular) which the Ministry of Supply was not fitted to carry. Thus it was decided to make over all work on atomic energy to a new, non-departmental authority. The United Kingdom Atomic Energy Authority came into being in 1954 and was made answerable to the Lord President of the Council. Under the Labour government the Minister of Technology acquired this responsibility.

In 1957 the National Institute for Research in Nuclear Science (NIRNS) was founded in order to provide cooperative research facilities for universities and other institutions; the investigation of sub-atomic physics is now largely inconceivable except in terms of ever more elaborate equipment and ever larger budgets to pay for it. Nevertheless the view has long been held by some that it takes a disproportionately large amount of the overall civil science budget to the detriment of other scientific fields.

The other big spender listed in the table above is the Ministry of Aviation, which was set up during the second world war as the Ministry of Aircraft Production, became the Ministry of Supply in 1946 (see above), the Ministry of Aviation in 1959, and in 1967 was submerged in the Ministry of Technology. Throughout, its bias has been military, acting as a procurement agency, carrying out aircraft and missile research and development, and being the aircraft industry's sponsor. In 1970 the Ministry was resurrected as the Ministry of Aviation Supply.

One of the major results of the second world war was that it made governmental intervention in science and technology respectable. Traditionally the Royal Society had acted as the government's informal advisory committee on matters affecting science policy. Throughout the interwar years the scientific expert assumed a greater importance, through agencies such as the Committee of Civil Research, than ever before. Scientific expertise at the highest level was further institutionalised after the war with the setting up of an Advisory Council on Scientific Policy and a Defence Research Policy Committee. The first only is of concern to us here since its interest lay in civil science and because of the long series of reports it and its subcommittees made until its winding up in 1965 and replacement by a new Council for Scientific Policy to oversee civil science, and an Advisory Council on Technology to assist the newly created post of Minister of Technology. These two bodies were linked through their common interest in manpower problems by a Committee on Manpower Resources for Science and Technology. Finally in 1966 the Central Advisory Council for Science and Technology was formed to assist the cabinet itself. The creation of such advisory bodies has been paralleled in many countries and they indicate a growing world-wide consciousness of the necessity for the evolution of national science policies, having implications – as we indicated by the definition quoted at the outset – far beyond the customarily assigned limits of science and technology.

Four Themes

The second part of the book, chapters 4 to 7, is devoted to four themes which there are treated separately although all have been touched upon in the first part.

Chapter 4 deals with research and development – science and technology in action – in the government's own establishments, in the Research Councils, the universities and that part of industrial research supported or financed by government. We touch upon the significance of pure as distinct from applied research, how it should be evaluated and priorities determined, the value of government research establishments, the Research Councils and the Research Associations and the scope for significant governmental intervention.

Chapter 5 relates to scientific manpower. The public debates over the appropriate numbers of qualified scientists and engineers, evident by the production of successive government projections of supply and demand, illustrate the importance of this subject as a separate study. Yet the idea of a 'manpower policy' in science and technology – or indeed, in any other professional area – is a comparatively new and hazardous example of central policy-making. It was only with the revelations of the early 1960s and the fears of a brain-drain of British scientists and engineers to the United States that the implications of such a policy were made evident. In many ways the confusion over the future production of trained men reflects the difficulty in predicting the future role of science and technology in society. The difficulties in manpower planning are not unique to Britain. They are reflected in every country in the world, irrespective of its degree of industrial and technical development. In the course of the chapter we have sought merely to illustrate, by reference to the successive reports which have appeared since the second world war, some of the commonly held assumptions and some of the inherent difficulties in planning for future manpower supply.

Chapter 6 concerns science in the Civil Service. The chapter has studied two main problems. First, the evolution of the Civil Service from an enclave of gifted amateurs to a highly competitive and expert body; second, the nature of scientific expertise and scientific advice. The first problem illustrates the scientist's institutionalisation within the Civil Service – the search for a role and the efforts made to integrate the scientist into the traditions of the Service. The second reflects the growing contribution of the scientist to decision-making in central government.

Chapter 7 concerns science and Parliament. This topic has been isolated for particular study because it presents a central dilemma in our conception of what science policy is and who is responsible for making it. It was only at the end of the nineteenth century that the gulf between the man of science and the politician became apparent. The increasingly complex and technically based legislation entailed by the necessity for controlling the ravages of industrialisation demanded that those who passed laws should understand the implications of their decisions. The gap between technical precision and legal control, manifest in the definition of the 'best practical means' for controlling such nuisances as air and water pollution, is but one example of the technical challenges which faced politicians of the time. With rare exceptions, few scientists either in this century or the last have been attracted to politics, yet there is an increasing amount of technical legislation reflecting the speed and the hazards of technological change. This chapter studies some of the ways in which Parliament, as an institution, has responded to these challenges.

Conclusion

We have tried to show the growing relationship of scientists with the state. With the benefit of hindsight we can see that the argument that the scientist must remain free of the state was doomed to failure (the enlistment of science and scientists in the world wars saw to that), but with the growth of the advisory and administrative machinery entailed by the adoption of science planning we can also see that some of the things feared by the Society for Freedom in Science (see Chapter 2) have come about. Thus the scientist's access to the politicians and the bureaucracy is limited. In other words, much science-policy making is 'closed' in nature. That political process in Britain is frequently like this is usually taken for granted and held to be a commendable quality of the system; however, it is alien to the spirit of science which has long cherished the 'openness' of its processes. J. W. Mitchell has expressed himself forcefully on this issue:[31]

No one disputes that councils, senates and administrations of

[31] *Chemistry and Industry*, 29 May 1965, p. 934.

universities, boards of directors of industrial firms and the Government and Parliament have the responsibility for determining the policies which affect scientists, and that once these policies are determined, it becomes the duty of the scientists to carry them out to the best of their abilities. What causes disquiet is the absence of any accepted procedure which allows the active leaders and the representatives of the scientists and their professional societies to be brought into consultations and given an opportunity of expressing their views before final and irrevocable decisions are made. . . . Secret and confidential operations are vital for national security in wartime and the people will then carry out the decisions of the Executive without hesitation. The processes of government and administration which allow the successful prosecution of a war should, however, not be applied extensively in times of peace, particularly to matters where no question of national security could possibly arise.

In the half decade since Mitchell wrote these words there have been a few small steps taken in the direction of greater openness: Parliament has established a Select Committee on Science and Technology which is able to probe matters of current concern (although the Committee, like all Parliamentary Committees, has no permanency, unlike American Congressional Committees), the concept of 'Green Papers' has been evolved – statements of government thinking intended for discussion and not as clear indications of proposed policy, and finally over one issue at least – whether Britain should support the CERN 300 GeV accelerator project – the government of the day published the evidence it had received.

The counterpart to this desire for more frankness on the part of government is that the scientific community should itself more clearly articulate how it sees its role and responsibility. But it is probably no more realistic to expect the corporate body of scientists and engineers to speak with one voice than to expect the community at large to do so. It is a matter of speculation whether the acceleration of scientific and technological development will compel the scientist to take a more active part in the political arena than he has done hitherto. A growing emphasis on international collaboration and cooperation, in the developed and the developing world, may have a profound effect on the nature of the scientific community, both in the rate and direction of research, and in the increasingly complex relationships between governments and scientists.

In the course of this book we have tried to illustrate some of the continuities and discontinuities apparent in the organisation of British science. The past may not hold the key to future patterns of development, but it is important that we understand those premises and assumptions on which we now base our science policy.

PART ONE
Historical Background

1 Central Organisation 1875–1919

The creation of the Advisory Council for Scientific and Industrial Research marked the culmination of over fifty years of argument and persuasion on the part of both scientists and statesmen for the consistent support of scientific research on a significant scale. Throughout the nineteenth century, scientific spending had been most conspicuous in the support of astronomical and geographical expeditions, in grants to individual scientists through the Royal Society and the British Association for the Advancement of Science, in piecemeal awards for scientific excellence through the Civil List and the medal system of the Royal Society, and in the strict utilitarianism of research done under the aegis of such departments as the Geological Survey, the Local Government Board and the Board of Trade. The Devonshire Commission of 1871–5 was unique in that it brought together those critics of government policy who argued for a more consistent policy for scientific research, and for the recognition of the scientific worker as a productive member of the community [1.1.1]. The value of the Commission's recommendations was somewhat impaired by the fact that the scientific community was neither unified nor organised in its demands for state support. The traditional organ of scientific advice to the government, the Royal Society, itself declined to assume the role of a pressure group, and this was largely left to those scientists writing in *Nature* under the enthusiastic leadership of Norman (later Sir Norman) Lockyer [1.1.3]. But not all scientists welcomed the prospect of increased public support for science. There were those who saw in it the threat of dependence upon the state for a livelihood and the distortion of scientific research into a tool of government policy [1.1.2]. It was not until after the turn of the century when, concerned with the Boer War and preoccupied with the concept of 'National Efficiency', that comparisons with Germany and the contribution of scientific research to the strength of the state began to make an appreciable impact on the public consciousness. The foundation of the National Physical Laboratory in 1899

[1.2.1] and the creation of the Development Commission (1909) and the Medical Research Committee (1913) were evidence of tentative steps towards public recognition of the intrinsic value of scientific research applied to national and social goals [1.2.2]. But these events were not evidence of a coherent policy for science either based on general popular scientific education or directed towards the recognition of scientific research as an essential aid to industrial development. It was not until the outbreak of the first world war that the shortages of trained men and key materials revealed the weakness of the government's attitude towards scientific research. The creation in 1915 of the Advisory Council for Scientific and Industrial Research [1.3.1–1.3.5], followed by the creation of the Imperial Trust [1.3.8] satisfied some of the demands of scientific pressure groups such as the British Science Guild. There were still, however, the associated problems of the place of science in national education and the status of the scientific worker. The first problem came under the scrutiny of the Thomson Committee in 1916 [1.4.2] and the second received welcome encouragement through the formation of the National Union of Scientific Workers in 1918.

The reappraisal of the machinery of government, conducted by the Haldane Committee of the Ministry of Reconstruction [1.5.1], confirmed the virtues of placing the Department of Scientific and Industrial Research under a Committee of the Privy Council. The Report recommended that similar arrangements be made for both medical and geological research, and in 1920 the Medical Research Committee became the second research council under a Committee of the Privy Council [1.5.3]. The Haldane Committee saw the research council formula as a necessary step towards the eventual formation of a Ministry of Research which would unite the research functions of government [1.5.1]; this view received strong support from H. A. L. Fisher, President of the Board of Education [1.5.2].

The period between 1900 and 1920 marked the origins of the organisation of scientific research as we now understand it. The search for 'policy' was undertaken in a negative way, in response to immediate problems rather than in a spirit of the imaginative use of resources. The movement for state support for science was largely carried by a small group of determined scientists and statesmen, and the methods chosen reflected the idealism of administrators and scientific advisers such as Frank Heath, William McCormick and Christopher Addison.

1.1 SOME PROPOSALS FOR REFORM, 1875–1903

1.1.1 The Devonshire Commission and the Concept of a 'Ministry of Science'

From the Eighth Report of the Royal Commission on Scientific Instruction and the Advancement of Science, under the chairmanship of the seventh Duke of Devonshire, C. 1298, 1875, pp. 45–6.

In 1869 Colonel Alexander Strange, FRS, succeeded in persuading the British Association for the Advancement of Science to appoint a committee for the specific purpose of obtaining a Royal Commission to study the role of government in supporting scientific research. They were successful, and the Royal Commission on Scientific Instruction, with the seventh Duke of Devonshire* as chairman and Norman Lockyer, editor of *Nature*, as secretary, began work in May 1870. Most scientists who gave evidence agreed that there was a need for a Science Ministry of some kind. Opinions differed as to what such a Ministry should do. Sir Richard Owen believed that there should be a minister for the administration of science and higher scientific instruction, advised by a scientific council. General Strachey, engineer and meteorologist, urged that a Scientific Council should be appointive, and held accountable for all its decisions. Captain Douglas Galton suggested an administrative council under the Privy Council. In a series of eight reports the Commission endorsed the view that scientific education needed encouraging; that state research institutions should be set up and research grants given; and that a special Council of Science under a Minister of Science should be established. The recommendations of the report had to wait nearly thirty years before their partial implementation.†

REMARKS ON THE FOREGOING EVIDENCE RELATING TO THE ESTABLISHMENT OF A MINISTRY AND COUNCIL OF SCIENCE
We have given careful consideration to this part of the Inquiry entrusted to us; and, in the course of our deliberations, we have been led to attach much importance to the facts stated in the First Part of our Report, which show that the Scientific Work of the Government is at present carried on by many different Departments.

* Short biographies of the Duke of Devonshire and Strange appear in J. G. Crowther's *Statesmen of Science* (Cresset Press, London, 1965).
† See R. V. Jones, *New Scientist*, 4 March 1971, pp. 481–3.

There is nothing to prevent analogous, if not actually identical, investigations being made in each of these, or to secure to one department an adequate knowledge of the results obtained, and the circumstances under which they were obtained, by another.

Investigations admitted to be desirable, nay, practical questions, the solution of which is of the greatest importance to the public administration, are stated by the witnesses to be set aside because there is no recognised machinery for dealing with them; while, in other cases, investigations are conducted in such a manner as to involve a needless outlay of time and money, because they were originally planned without consultation with competent men of Science.

Passing to the question of the Advancement of Science, we have arrived at the conclusion that much has to be done which will require continuous efforts on the part of the Administration unless we are content to fall behind other Nations in the Encouragement which we give to Pure Science, and, as a consequence, to incur the danger of losing our pre-eminence in regard to its Applications.

These considerations, together with others which have come before us in the course of our Inquiry, have impressed upon us the conviction that the Creation of a Special Ministry dealing with Science and with Education is a Necessity of the Public Service.

This Ministry would be occupied (1) with all questions relating to Scientific and General Education, so far as these come under the notice of Government; (2) with all questions incidental to the application of National Funds for the Advancement of Science; and (3) with all Scientific Problems in the Solution of which the other Departments may desire external Scientific Advice or Information. It would also be desirable that the Department should receive Information as to Scientific Investigations proposed by other branches of the Government, and record their progress and results.

It is not within our province to express an Opinion as to whether the subject of Art should be included among the Functions of this Department; but we are satisfied that the Minister's attention should not be distracted by any immediate Responsibility for affairs which have no connexion with Science, Education, or Art.

We have considered whether the Official Staff of such a Ministry, however carefully selected, could be expected to deal satisfactorily with all the varied and complicated Questions which would come before the Department. We have given full weight to the Objections which have been raised against the creation of a Special Council of Science, and to the Arguments in favour of referring Scientific Questions to Learned Societies, or to Special Committees appointed for the purpose, or to private Individuals; but nevertheless we have arrived at the Conclusion that an additional Organisation is required through which the Minister of Science may obtain Advice on questions involving Scientific considerations, whether arising in his own Department or referred to him by other Departments of the Government.

Such questions have from time to time been referred to the Council of the Royal Society, in which the best Scientific Knowledge of the time is fairly represented. The Committee chosen by that Council for the Administration of the Government Grant of 1,000*l.* per annum in Aid of Scientific Investigations has performed its work to the satisfaction of the Government, of men of Science, and of the Public. But if much more is to be done for the Advancement of Science than at present, and if the Departments in conducting their Investigations are to have the benefit of the Scientific Advice which appears now to be frequently wanting, the Council of the Royal Society, chosen as it is for other purposes, could scarcely be expected to take upon itself Functions which, it is true, are not different in kind, but which would involve increased Responsibility and the Expenditure of additional time and trouble. Moreover, amongst the questions on which the Departments would require Scientific Advice, there would no doubt be many requiring a Knowledge of the peculiar exigencies of the Public Service, which would be more readily understood and solved if some persons in direct relation with the Departments formed a part of the body to be consulted. It is obviously of great importance that the Council should be so constituted as to possess the confidence of the Scientific World, and we believe that this confidence would be extended to a Council composed of men of Science selected by the Council of the Royal Society, together with Representatives of other important Scientific Societies in the United Kingdom, and a certain number of persons nominated by the Government. We also believe that such a Body would deserve and receive the confidence of the Government; and that it would be well qualified to administer Grants for the Promotion of Pure Science.

The general opinion we have expressed as to the proper

31

Remuneration of Scientific Work would be applicable to the Members of this Council, but the degree and manner in which the principle should be applied in this instance must be so largely dependent on circumstances that we cannot make any specific Recommendation on the subject.

It would be impossible that the Council should in all cases undertake the direct solution, by itself or even by Sub-Committees, of the problems submitted to it. In many instances, especially when experimental investigations are required, its duty would be accurately to define the problem to be solved, and to advise the Minister as to the proper persons to be charged with the Investigation.

We are of opinion that the Council should not have the power of initiating Investigations; it should, however, not be precluded, in exceptional cases, from offering to the Minister such suggestions as it may have occasion to make in the Public Interest.

We believe that Reference to such a Council would be found to be so useful and convenient that it would become the usual course in cases of difficulty, but we would not diminish the Responsibility or fetter the Discretion of any Minister by making such Reference obligatory, or by preventing a Reference to Committees or to Individuals chosen by him, whenever that course might appear to him to be more desirable.

1.1.2 The Dilemma of the State-supported Scientist

From Michael Foster, 'The State and Scientific Research', *Nineteenth Century*, May 1904, vol. 55, pp. 742–5.

Professor Michael (later Sir Michael) Foster was Professor of Physiology at Cambridge from 1883 to 1907. From 1900 to 1906 he was the Liberal Unionist MP for London University. In the following article he expressed the dilemma of the state-supported scientist. The independent scientist, he argued, was free to choose his research and pursue his imagination; the 'state scientist' was bound by economic dependence to the utilitarian goals of the state. His solution was that the state should contribute substantially to universities and scientific research throughout the country but should not create its own laboratories. (An implicit criticism of the National Physical Laboratory [cf. 1.2.1].) Thus both the independence of the scientist and the value of research to the community could be maintained.

The spirit which rules the State in its ordinary payment of scientific work is, put baldly, that it should have its money's worth in return for the money spent. Let us suppose that in preparing for administrative action or for legislative measures a scientific question presents itself, and that an answer to this question must be found before the action is begun or the measure framed. The money value of the answer will depend on the importance of the action or measure itself, and on the prominence of the answer in shaping the action or framing the measure. The State can, by the help of these two factors, determine in a business way, within certain limits, how much it is justified in paying for the scientific work which provides the answer. It has itself no interest at all, or a very indirect one only, in the scientific features of the answer; it accepts and must accept the answer, whatever it be, as the judgement of science at the present day, and shapes its action or frames its measures accordingly. If the answer be clear and sharp, so much the better; if, as frequently happens, the question put being a difficult and complex one from a scientific point of view, the answer is vague and uncertain, it still answers its purpose, and can be paid for without offence to the business conscience; the only difference is that the action has to be shaped or the measure framed in a way to suit the vagueness and uncertainty.

In accordance with this spirit the paid scientific servant of the State, who for hire is working out the answer, must keep his eyes steadily fixed on the answer, and on that alone; he must not wander away from the prescribed end; he must not be tempted to follow up questions which incidentally arise in the inquiry, but the answers to which might and probably would have no bearing on the main issue; a narrow path is laid out for him, he must not stray from it.

Very different is the spirit which guides the independent man of science, the private worker in any inquiry which he takes up; for that spirit is the spirit of perfect freedom. In the first place he is free to choose his inquiry; he selects a problem, not because that particular problem is allotted to him by someone else, but because the problem attracts him. Something within him pushes him on to solve it; or rather – and this is the happier case – the problem is such as to give him no rest until he has at least done his best to solve it. He starts on the inquiry with an impulse denied to him who undertakes a task not chosen by himself but offered to him by others.

In the second place – and this is the really important point – having taken the first steps towards the solution of the problem, having put the first questions, he is free to follow wherever the initial answers seem to beckon him. And experience has shown that absolute freedom to follow wherever Nature leads is the one thing needful to make an inquiry a truly fruitful one. Whatever be the nature of the inquiry on which a man sets out, whatever be the initial problem which starts him on his quest, sooner or later, as his work unfolds before him, there comes a time when there stares him in the face some new question arising out of the results which he is gathering in, but leading him right away from the goal immediately before him. The new question may be a small thing; if so, he merely notes it and passes on. But it may hold within itself the promise of things far greater than any possible result of the inquiry in hand. What is he in this case to do? Shall he, abandoning the old inquiry, throw himself heart and soul into the new inquiry thus suddenly opened up? Occasions such as these test the possession of the real gift of inquiry, the genius of research. The man who has not this gift either loses his way in fruitless wanderings into barren side issues, or misses golden opportunities by refusing to pay heed to any side issue. He who has the gift seizes on the right side issue, giving up everything to follow that; and the story of science shows us how again and again a new truth, standing out as a great landmark, has been won by the man of genius who dared to leave his main inquiry and follow out the trail which crossed his path.

In the third place, the private independent inquirer is free to be what we may call reckless in his research; and this is no slight advantage. Charles Darwin in the course of his work was again and again led to make what he used to speak of as a 'fool's experiment', an arrow shot without definite aim into the unknown in the hope that it might bring something down. And we know from his Life that, not once only, a fool's experiment became in his hands the starting point of a new line of fruitful inquiry. Freedom thus to play 'the fool' is no mean help towards success in research.

All these several advantages, born of perfect freedom, are denied to the State-aided inquirer.

He takes up this inquiry rather than that, not because the spirit calls him, but because the State bids him. In the inquiry he must be ever sober, never reckless; the cost of a 'fool's experiment' must not appear in the estimates laid before the public. And, above all, he cannot follow side issues. He is bound by contract to bring the research on which he has been put to its own proper end. Trail after trail may cross his path; he may have visions of the splendid scenes to which this or that cross trail might lead him, did he dare to follow it up; but he must not step aside, he must push straight on towards the goal to reach which he is paid. He may tell others of this or that side issue, and they may reap the success denied to him. Or, keeping it to himself, he may mark it in his notebook, meaning to return to it when his main work is done. But he will never return to it. He will forget it; or new work imposed on him will prevent his touching it; or if, haply, the opportunity to take it up again should come to him, he will find that the inspiration which moved him when the thought first came to him has vanished, the virtue has gone out of him, the accepted time has passed; he can now do nothing with the idea which once seemed so full of promise.

These are some of the reasons which bring it about that, while a poor man's mite invested in a free, untrammelled, perhaps even reckless, inquiry may come back again as a new scientific truth, not only of great theoretic beauty, but of great practical value, the contributions to science which appear as the return for the large sums spent by the State on official scientific investigations are rarely, if ever, more than mediocre and are sometimes even poor. Research which the Government orders and pays for by the piece may perform some of the drudgery of science, may produce good meritorious useful work, though the usefulness is often curtailed by reason of the narrow limits imposed on the inquiry; it does not and cannot be expected to give birth to new creative ideas, to thoughts which move mankind. Yet these are the results of real value, even of real money value. The State may think that it gets its money's worth for its money, because it has received an answer to the question which it put, and feels assured that for any vagueness and uncertainty in the answer, not the inquirer whom it pays but the present state of science is to blame. How much more than its money's worth might it have got had the same money been spent in an inquiry the result of which wholly changed the present state of some branch of science!

We are placed in this dilemma: inquiry needs money, but has even greater need of freedom; the State can offer money, but it gives money at the cost of freedom. Is there no way out of the dilemma? Is there no plan by which the State can spend money on inquiry, and secure at one and the same time that inquiry should on the one hand enjoy that perfect freedom which is essential to its being fruitful, to its bringing forth new ideas and making real advance, and on the other hand be always prepared to furnish answers to questions raised by the needs of the State?

1.1.3 The need for a National Scientific Council: the Origins of the British Science Guild.

From Sir Norman Lockyer, Presidential Address to the British Association, 1903, 'The Influence of Brain Power on History', British Association Annual Report for 1903, pp. 9, 26–8.

Sir Norman Lockyer's Presidential Address was partly in response to those who saw national resources only in terms of military armaments. His message was simple: British statesmen had neglected to organise or use the nation's intellectual resources and British industry had turned its back on scientific innovation. The superiority of the Germans in science and technology was undeniable. Lockyer's remedy was, first, to increase endowments to British universities for scientific teaching and research and to make them comparable to both German and American universities; second, the creation of a National Scientific Council as the Devonshire Commission had recommended; and third, that scientists should organise themselves into a chamber, guild or league, in order to press their interests upon government. Lockyer's address had a wide popular appeal, but neither the British Association nor the Royal Society took up his plea and the idea of a National Scientific Council was neglected for another ten years. In 1905, therefore, Lockyer set up his own pressure group, the British Science Guild, to bring scientific education, teaching and research to public notice.

The Necessity for a Body dealing with the Organisation of Science
The present awakening in relation to the nation's real needs is largely due to the warnings of men of science. But Mr Balfour's terrible Manchester picture of our present educational condition shows that the warning, which has been going on now for more than fifty years, has not been forcible enough; but if my contention that other reorganisations besides that of our education are needed is well founded, and if men of science are to act the part of good

citizens in taking their share in endeavouring to bring about a better state of things, the question arises, Has the neglect of their warnings so far been due to the way in which these have been given?

Lord Rosebery, in the address to a Chamber of Commerce from which I have already quoted, expressed his opinion that such bodies do not exercise so much influence as might be expected of them. But if commercial men do not use all the power their organisation provides, do they not by having built up such an organisation put us students of science to shame, who are still the most disorganised members of the community?

Here, in my opinion, we have the real reason why the scientific needs of the nation fail to command the attention either of the public or of successive Governments. At present, appeals on this or on that behalf are the appeals of individuals; science has no collective voice on the larger national questions; there is no organised body which formulates her demands.

During many years it has been part of my duty to consider such matters, and I have been driven to the conclusion that our great crying need is to bring about an organisation of men of science and all interested in science similar to those which prove so effective in other branches of human activity. For the last few years I have dreamt of a Chamber, Guild, League, call it what you will, with a wide and large membership, which should give us what, in my opinion, is so urgently needed. Quite recently I sketched out such an organisation, but what was my astonishment to find that I had been forestalled, and by the founders of the British Association!

The Need of a Scientific National Council

In referring to the new struggle for existence among civilised communities I pointed out that the solution of a large number of scientific problems is now daily required for the State service, and that in this and other ways the source and standard of national efficiency have been greatly changed.

Much evidence bearing upon the amount of scientific knowledge required for the proper administration of the public departments, and the amount of scientific work done by and for the nation, was brought before the Royal Commission on Science presided over by the late Duke of Devonshire now more than a quarter of a century ago.

The Commission unanimously recommended that the State should be aided by a scientific council in facing the new problems constantly arising.

But while the home Government has apparently made up its mind to neglect the advice so seriously given, it should be a source of gratification to us all to know that the application of the resources of modern science to the economic, industrial, and agricultural development of India has for many years engaged the earnest attention of the Government of that country. The Famine Commissioners of 1878 laid much stress on the institution of scientific inquiry and experiment designed to lead to the gradual increase of the food-supply and to the greater stability of agricultural outturn, while the experience of recent years has indicated the increasing importance of the study of the economic products and mineral-bearing tracts.

Lord Curzon has recently ordered the heads of the various scientific departments to form a board, which shall meet twice annually, to begin with, to formulate a programme and to review past work. The board is also to act as an advisory committee to the Government, providing among other matters for the proper co-ordination of all matters of scientific inquiry affecting India's welfare.

Lord Curzon is to be warmly congratulated upon the step he has taken, which is certain to bring benefit to our great Dependency.

The importance of such a board is many times greater at home, with so many external as well as internal interests to look after – problems common to peace and war, problems requiring the help of the economic as well as of the physical sciences.

It may be asked, What is done in Germany, where science is fostered and utilised far more than here?

The answer is, There is such a council. I fancy, very much like what our Privy Council once was. It consists of representatives of the Ministry, the Universities, the industries, and agriculture. It is small, consisting of about a dozen members, consultative, and it reports direct to the Emperor. It does for industrial war what military and so-called defence councils do for national armaments; it considers everything relating to the use of brain-power in peace – from alterations in school regulations and the organisation of the Universities, to railway rates and fiscal schemes, including the adjustment of duties. I am informed that what this council advises, generally becomes law.

It should be pretty obvious that a nation so provided must have enormous chances in its favour. It is a question of drilled battalions against an undisciplined army, of the use of the scientific spirit as opposed to the hope of 'muddling through'.

Mr Haldane has recently reminded us that 'the weapons which science places in the hands of those who engage in great rivalries of commerce leave those who are without them, however brave, as badly off as were the dervishes of Omdurman against the maxims of Lord Kitchener'.

Without such a machinery as this, how can our Ministers and our rulers be kept completely informed on a thousand things of vital importance? Why should our position and requirements as an industrial and thinking nation receive less attention from the authorities than the headdress of the Guards? How, in the words of Lord Curzon, can 'the life and vigour of a nation be summed up before the world in the person of its sovereign' if the national organisation is so defective that it has no means of keeping the head of the State informed on things touching the most vital and lasting interests of the country? We seem to be still in the Palaeolithic Age in such matters, the chief difference being that the sword has replaced the flint implement.

Some may say that it is contrary to our habit to expect the Government to interest itself too much or to spend money on matters relating to peace; that war dangers are the only ones to be met or to be studied.

But this view leaves science and the progress of science out of the question. Every scientific advance is now, and will in the future be more and more, applied to war. It is no longer a question of an armed force with scientific corps; it is a question of an armed force scientific from top to bottom. Thank God the Navy has already found this out. Science will ultimately rule all the operations both of peace and war, and therefore the industrial and the fighting population must both have a large common ground of education. Already it is not looking too far ahead to see that in a perfect State there will be a double use of each citizen – a peace use and a war use; and the more science advances, the more the old difference between the peaceful citizen and the man at arms will disappear. The barrack, if it still exists, and the workshop will be assimilated;

the land unit, like the battleship, will become a school of applied science, self-contained, in which the officers will be the efficient teachers.

I do not think it is yet recognised how much the problem of national defence has thus become associated with that with which we are now chiefly concerned.

These, then, are some of the reasons which compel me to point out that a scientific council, which might be a scientific committee of the Privy Council, in dealing primarily with the national needs in times of peace, would be a source of strength to the nation.

1.2 STEPS TOWARDS THE STATE SUPPORT OF RESEARCH

1.2.1 Early Days at the National Physical Laboratory

> From Sir Richard Glazebrook, FRS, a lecture delivered at the National Physical Laboratory on 23 March 1933.

> Glazebrook was the first director (1900–19); in this lecture he outlined the events leading to the setting up of the Laboratory in 1899.

In order to understand the development of the National Physical Laboratory during the early period of its activities it is necessary to go back some years before its actual foundation. Its birth-throes were long and tedious, its teething troubles many, and its survival was due to the unstinted help and generosity of a number of friends, the constant support of the Royal Society, and the devoted work of a small staff.

The years that followed the conclusion of the Franco-German War saw a great expansion of industry in Germany, an expansion supported by von Helmholtz and Werner von Siemens on the scientific side. They knew that science could give help of the greatest value to technical endeavours, and they pointed out the methods by which this could be done. The result was the establishment of the Physikalisch-Technische Reichsanstalt at Charlottenburg, a suburb of Berlin, during the period 1883–7. Von Helmholtz became its first Director.

In 1891 the British Association met at Cardiff, and in his Presidential address to Section 'A', Sir Oliver Lodge called attention to the work in progress at the Reichsanstalt and the importance to

British industry of a similar Institution. Nothing could be done without Government help, and in the political circumstances at that time to secure this seemed impossible. A Committee was appointed as the result of the Presidential address, but lapsed without taking further practical action.

In that address Sir Oliver said, after a reference to the work at Kew and in the Board of Trade Laboratory,

But what I want to see is a much larger establishment erected in the most suitable site, limited by no speciality of aim nor by the demands of the commercial world, furnished with all appropriate appliances, to be amended and added to as time goes on and experience grows, invested with all the dignity and permanence of a national institution, a physical laboratory in fact comparable with Greenwich Observatory and aiming at the very highest quantitative work in all departments of physical science.

May we claim forty years afterwards that this has been achieved?

In 1895, Sir Douglas Galton was President at the Ipswich meeting of the Association. He raised the matter in his Presidential address and developed it in a paper read before Section 'A'. Let me quote from his address:

There could scarcely be a more advantageous addition to the assistance which Government gives to science than for it to allot a substantial annual sum to the extension of the Kew Observatory in order to develop it on the model of the Reichsanstalt.

It might advantageously retain its connexion with the Royal Society under a Committee of Management representative of the various branches of science concerned and of all parts of Great Britain.

In his paper before Section 'A' the President suggested the resuscitation of the 1891 Committee to report

(a) upon the functions which an establishment of this nature should fulfil;

(b) upon the system which should be adopted for its control and management.

As the result a Committee* was appointed to consider:

* The members were: Sir D. Galton (Chairman), Prof. O Lodge (Secretary), Lord Rayleigh, Lord Kelvin, Sir H. Roscoe, Professors Rücker, Clifton, Carey Foster, Schuster, Ayrton, Dr Anderson, Dr T. E. Thorpe, Mr Francis Galton, Mr Glazebrook.

the establishment of a National Physical Laboratory for the more accurate determination of Physical Constants and for other quantitative research, and to confer with the Council.

It will be noted, and is of interest, that Standardisation and Testing are not mentioned in the reference to the Committee. The Committee reported to the meeting at Liverpool in 1896 in favour of the establishment of such a laboratory. They recommended that the definition of its functions and its management should be in the hands of the Royal Society, but that 'the immediate executive and initiative power should rest in a paid chief or Director'. The Report continues: 'We recommend that an annual grant in addition to the sum already expended' (*i.e.* at the Kew Observatory), 'of £5,000 per annum, and an additional initial expenditure of £30,000 for building and equipment, would do all that is essential to carry out the scheme in a wise and worthy manner.'

The history is continued in the opening words of Professor Forsyth's address as President of Section 'A' at Toronto in 1897. After a reference to the action taken in previous years, he said:

Thereupon a joint committee representing the various scientific bodies throughout the United Kingdom interested was constituted to further the plan. . . .

It was a deputation from this joint committee which, headed by Lord Lister, waited on the Prime Minister (Lord Salisbury) on February 16th last. His reply to the deputation was manifestly sympathetic with the request, and it is a satisfaction now to be able to say that the Government have appointed a committee of inquiry which will also consider whether standardisation and other work already undertaken partly or wholly at the public cost can fitly be associated with the new institution.

The British Association has done much for British science, its action in promoting a National Physical Laboratory is not the least of its many benefits.

The reference to the Treasury Committee was in the following terms:

To consider and report upon the desirability of establishing a National Physical Laboratory for the testing and verification of instruments for physical investigation, for the construction and preservation of standards of measurement and for the systematic determination of physical constants and numerical data useful for scientific and industrial purposes – and to report whether the work of such an institution, if established, could be associated with any testing or standardising work already performed whether wholly or partly at public cost.

It will be noticed that standardisation and testing occupy a prominent place in this reference. The Committee consisted of Lord Rayleigh (Chairman), Sir Courtenay Boyle (Secretary of the Board of Trade), Mr Chalmers (of the Treasury), Sir John Wolfe Barry, Sir Andrew Noble, Mr Siemens, and Professors Rücker, Thorpe and Roberts-Austen.*

The Committee reported on 6 July 1898, and their conclusions were:

1 That a public institution should be founded for standardising and verifying instruments, for testing materials, and for the determination of physical constants.

2 That the institution should be established by extending the Kew Observatory in the Old Deer Park, Richmond, and that the scheme should include the improvement of the existing buildings, and the erection of new buildings at some distance from the present Observatory.

3 That the Royal Society should be invited to control the proposed institution, and to nominate a Governing Body, on which commercial interests should be represented, the choice of the members of such Body not being confined to Fellows of the Society.

4 That the Permanent Secretary of the Board of Trade should be an *ex-officio* member of the Governing Body; and that such Body should be consulted by the Standards Office and the Electrical Standardising Department of the Board of Trade upon difficult questions that may arise from time to time or as to proposed modifications or developments.

The Report was sent to the Royal Society by Sir E. W. Hamilton, Secretary to the Treasury, in a letter dated 7 October 1898, from which the following clauses are taken:

In the first place, Her Majesty's Government are prepared to

* These gentlemen became at a later date members of the first Executive Committee of the National Physical Laboratory.

adopt generally the four conclusions of the Committee as set forth on p. 6 of the Report.

After a reference to the fact that Kew Observatory is managed by a Royal Society Committee, the letter continues:

It is therefore the duty of this Board to ask whether the Royal Society is prepared to concur in these conclusions of the Committee and to cooperate in carrying them into effect. But before answering this question the Royal Society will doubtless desire to know what pecuniary assistance may be expected from public funds towards the scheme recommended by the Committee.

Then after a reference to possible receipts from fees and from the Gassiot Fund—

Beyond these sources of income Her Majesty's Government would be prepared to ask Parliament to vote an annual sum of £4,000 for five years certain as a grant in aid of the expenses of the proposed institution. As regards buildings the existing Kew Observatory would remain available for its present or for similar uses, and, in addition, Parliament would be asked to make a grant not exceeding £12,000 in all towards the erection of suitable buildings.

This letter was presented to the Council of the Royal Society together with a memorandum by the Secretary, Professor Rücker, and referred to a Committee consisting of Professor W. G. Adams, Professor Clifton, Professor Ewing, Mr Kempe, Dr Larmor, Professor Meldola and Dr Russell (with the President and officers) to consider and report to the Council at its next meeting.

Correspondence with the Treasury followed, mainly as to details relating to Kew Observatory, and on December 1898, a letter from Sir F. Mowatt was read to the Council replying to the questions raised by the Committee and approving of certain suggested modifications. This was referred by the Council to the Committee with power to act in direct communication with the Treasury.

On 9 January 1899, a draft scheme was submitted to Council by the Committee and approved, subject to the consideration of further suggestions from the Kew Committee, should any be received; two months later (16 March 1899) a letter from the Treasury approving the scheme was laid before the Council.

Thus the Laboratory was started.

1.2.2 State Support for Medical Research

From the First Annual Report of the Medical Research Committee for 1914–15, Cd. 8101, 1914–16, pp. 5–8.

Following the creation of the Development Commission in 1909 the next stage was the formation in 1913 of the Medical Research Committee, under the auspices of the National Insurance Act of 1911. The Report for 1913–14 on the Administration of National Health Insurance described the origins of the Medical Research Committee in the recommendations of the Final Report of the Departmental Committee on Tuberculosis. The Departmental Committee had expressed the opinion that government could not make the best possible use of the money for research into tuberculosis provided in accordance with the provisions of the National Insurance Act at the rate of one penny per insured person without having the help and advice of experts in the field. Two bodies were subsequently set up; one consisting of forty-two persons, called the Advisory Council for Research and one, a smaller executive body of nine persons, more directly concerned with the allocation of the money. The Advisory Council for Research was made up of representatives of professional medical associations and state departments; the smaller Medical Research Committee reflected the recommendation of the Departmental Committee's report that the majority, but not all, of the members should be scientific experts. The following extract describes how the Medical Research Committee saw its first and immediate responsibilities.

The Medical Research Committee and Schemes of Research
The first Members of the Advisory Council and of the Medical Research Committee were formally appointed to serve for three years from 20 August 1913.

The Regulations lay down that after that period three members of the Committee, to be selected in such manner as the Committee may determine, shall retire at intervals of two years, but shall be eligible for reappointment. The members of the Council are eligible for reappointment after the period of three years.

The Medical Research Committee, under the chairmanship of Lord Moulton, began without delay to consider the organisation of the new State resources for the advancement of medical knowledge. During the summer and autumn of 1913 the Committee closely studied the numerous questions connected with the formation of a national scheme of medical research, and before proceeding to the choice and distribution of particular subjects for research to be initially undertaken they desired to seek approval for the general outline of the policy they were led to propose.

Their general scheme was submitted to the Minister in the following terms:

Type of Research.—The object of the research is the extension of medical knowledge with the view of increasing our powers of preserving health and preventing or combating disease. But otherwise than that this is to be the guiding aim, the actual field of research is not limited and is to be wide enough to include, so far as may from time to time be found desirable, all researches bearing on health and disease, whether or not such researches have any direct or immediate bearing on any particular disease or class of diseases provided that they are judged to be useful in promoting the attainment of the above object.

Method of conducting the research.—The organisation by which this research will be carried out should consist of the following departments:

1. A competent body of investigators of the highest class in the permanent employ of the scheme and devoting their whole time to research under it. They would be supplied with proper laboratories, duly qualified assistants, etc., and would ordinarily carry on their researches in such laboratories.

2. Skilled investigators in the permanent or temporary employment of the scheme who would be engaged in procuring their material clinically or otherwise in connexion with Hospitals or other Institutions furnishing the requisite opportunities for so doing. This material would in some cases be worked upon in local laboratories, and in some cases at laboratories provided for them elsewhere under the scheme, and sometimes by a combination of both methods.

3. Individual investigators not in the employment of the scheme who are carrying on independent investigations of a kind which are suitable to form a part of, or to be coordinated with, the research under the scheme, and to whom it is desirable to give help either in money or otherwise to enable them better to carry on their researches.

4. Statistical Department.—This will mainly consist of persons in the permanent employment of the scheme who will be engaged in inquiries relating to diet, occupation, habits of life and other matters bearing upon the incidence of disease, and who will collect and deal with all types of vital statistics, including the distribution of disease, the relative frequency of special types of lesions in diseases, such as tuberculosis and in general with all statistical investigations useful either as preliminary to research or as confirmatory of its results. It will probably have to consider and advise how the statistical material provided for under the Insurance Act should be dealt with.

It is hoped that when the scheme is in actual work there may become associated therewith a Bureau through which those engaged in research unconnected with the scheme or otherwise working on kindred questions may be able to obtain information, references to special publications and other help of a like nature.

All these four departments are essential to the success of the organisation and are intended to cooperate one with another and will be used separately or together according to the nature of the work in hand. It is neither possible nor desirable to lay down any hard and fast lines of demarcation of their spheres of action.

The Committee estimated roughly that £60,000 would be needed for the provision and equipment of the Central Research Institute, and proposed that this capital expenditure should be spread over two years, leaving in each of those years about £25,000, as to which they proposed that about £10,000 should be devoted to the expenses of administration, of the libraries and information bureau, and of research centralised in the Institute or immediately within the control of the Committee, while about £15,000 should be assigned to researches prosecuted by approved investigators at the Universities and other research centres maintained by funds voluntarily provided. After the first two years and the completion of the initial capital expenditure, larger amounts would be available for the expected growth in the current research work, both centrally and at local Institutions, as the schemes for research reached their full development.

47

It will be seen that this general plan for the future work of the Committee marks clearly a line of policy by which part, but only a part, of the new national fund for research is devoted to work centralised in laboratories under their control and carried out by a scientific staff to be directly appointed by the Committee, while the other and larger part of the fund is allocated to the support of workers and their investigations in laboratories and institutes not directly under the control of the Committee. It is a fundamental feature of this general scheme that independent investigations within the voluntary and private institutions in which almost the whole of the research work hitherto carried out in Great Britain, with little, if any, support from State authorities, should receive assistance from the Medical Research Fund, insofar as their work is appropriate to medical research, or capable of being coordinated within a general scheme for the organised advancement of medical knowledge.

The general principle here laid down and finally approved by the Minister, after a meeting of the Advisory Council held on 4 December 1913, is consonant with the recommendation of the Departmental Committee on Tuberculosis, who in remarking that, 'hitherto, apart from the small annual sum expended by the Local Government Board and actual grants for particular objects, the State has, in the main, left Research to voluntary agencies', express not only their high approval of the valuable work in medical research which has been rendered by voluntary effort in the past, but also their opinion that the aim of those responsible for the organisation of research under the National Insurance Act should be 'to stimulate and cooperate with the voluntary agencies'.

The Chairman of the Joint Committee, in signifying his approval of the scheme so outlined, gave it with the proviso that the sum (£60,000) estimated for capital expenditure should be spread over the first five instead of over the first two years, and this modification of the proposal of the Committee was communicated to the members of the Advisory Council on 19 December 1913, by a circular letter from Lord Moulton, as Chairman of the Council. At the same time, with a view to the establishment of a Central Research Institute, the provisional Regulations made by the Joint Committee and dated 20 August 1913, were modified so as to allow the Medical Research Committee to acquire land or interest in it, or erect buildings and provide their maintenance, and this modification appeared in the substantive Regulations dated 21 March 1914, which now govern the work of the Committee.

1.3 SCIENCE IN WARTIME: THE CREATION OF THE DEPARTMENT OF SCIENTIFIC AND INDUSTRIAL RESEARCH

1.3.1 Science, Education and Industry: the First Proposals

From Board of Education, Proposals for a National Scheme of Advanced Instruction and Research in Science, Technology and Commerce, April 1915, Public Record Office Ed. 24/1581.

The following proposals were devised mainly by Dr Christopher (later Sir Christopher) Addison, Parliamentary Secretary to the Board of Education, Sir Amherst Selby-Bigge, Permanent Secretary of the Board of Education, and Sir Frank Heath. They envisaged a reorganisation, not only of the relationship between research and industry, but also of secondary school science and undergraduate science in universities. These plans were later reshaped, due to the immediate pressures and exigencies of war, into a narrower programme dealing mainly with industrial and university research. A valuable opportunity to relate scientific teaching, research and industry from secondary school to the final stages of industrial production was thereby lost.

1. My colleagues will have realised how greatly many of our industries have suffered during the war through our inability to produce at home certain articles and materials required in trade processes, the manufacture of which has become localised in Germany through the superior skill and energy which the Germans have applied to the organisation of scientific research. On the conclusion of peace the Germans will devote the same power of scientific work and organisation but with renewed energy to the task of capturing our markets, both at home and abroad. It is unnecessary for me to emphasise the grave danger, possibly amounting to economic ruin, which will be incurred by our industries if they have to face fresh and embittered competition without enlarged scientific resources. The advantages, sometimes amounting to a practical monopoly, which the Germans have secured over our manufactures in dyes,

drugs, scientific glass making, X-ray instruments, photographic paper, and many other branches of industry has [*sic*] been due to the scarcity of scientific research work in this country and the small use made of it by industrial firms, rather than to protection. In the United Kingdom, for instance, there are only 1,500 chemists, including analysts, employed in the whole of our industries. In Germany the four chief firms in the one industry of colour work alone employ 1,000 research chemists. Similar relations hold good in many other of our industries, and it appears to me to be incontrovertible that if we are to maintain our industrial position we must, as a nation, aim at such a development of scientific, technological, and commercial education, and research as will place us in a position to compete successfully with the most highly organised of our rivals.

2. The problem is twofold. We have not merely to make the best use of the scientific men we now possess, but to provide a fuller supply in the future. The deficiency in the supply is mainly due to the slight prospect which applied science offers of a useful and remunerative career, and this in turn is due to the fact that the leaders of industry do not appreciate the service which science might render to them, partly because the training in our universities has hitherto been directed to the requirements of examinations rather than to the application of knowledge to industry, and partly because the facilities for postgraduate research into problems of direct concern to industry have hitherto been very inadequate. Further provision for postgraduate work is, therefore, essential, but we have at the same time to build from the foundations an educational ladder by which the ablest boys in the country may climb through the secondary schools or technical colleges up to the universities. The pick of those who have completed their university course will either become industrial experts or researchers, and, as regards the latter, we must provide some organisation for securing the cooperation of researchers in the same branch of work.

3. During the last few weeks, Dr Addison and my staff have been giving very careful consideration to the details of the problem, and as a result of their investigation, in consultation with experts, the following scheme is recommended.

4. The first part of our scheme aims at sifting out and assisting:

(1) Those pupils whose capacity justifies public expenditure in continued training;

(2) Those who should be assisted to receive whole-time instruction in higher technical institutes or universities;

(3) Those who, on the completion of their university course, are fit to receive scholarships or fellowships in industrial research.

At the same time my proposals provide for increased grants to the several kinds of schools or colleges to enable them to deal with increasing efficiency with the increasing number of pupils who may be expected to attend them.

(c.) Universities and Higher Technical Institutes: Whole Time

(i) Scholarships.—All the arguments for lengthening the school life of the best elementary school children apply to extending the further education of the best pupils in technical classes. The economic difficulty is here even more strongly felt, and State aid is, therefore, required on a more generous scale. I ask for 11,000*l.* in the year 1916–17 to be expended on scholarships tenable at universities and higher technical institutes by 120 whole-time students, rising afterwards to 300. The scholarships, which would be of an average value of 90*l.*, would be mainly distributed between the engineering, building, and woodwork trades, chemical industries, mining, textiles, commercial and pure science subjects.

I also ask for 15,000*l.* to be expended in scholarships of the value of 90*l.*, tenable at similar institutions by carefully selected students from secondary schools.

(ii) Grant in aid of Fees.—In addition to the financial aid given by way of scholarships, my proposals include a uniform degree course fee of 10*l.* a year in accordance with the recommendations of Lord Haldane's Commission. To make good the loss to the universities we should require an additional grant of 66,000*l.* in 1916–17.

(iii) Grant for Maintenance and Equipment.—At the expiry of the present quinquennium in 1916 I shall ask for an additional 100,000*l.* to enable the universities in receipt of grant to improve their staff and equipment and provide facilities for research. In addition to the direct advantages to industry this will give men

51

who have hitherto turned to other callings a reasonable prospect of a career within the university itself.

5. The second part of our scheme is directed to the promotion of research work into industrial and commercial problems. For this purpose I am asking my colleagues for sanction to spend in 1916–17 the sum of 90,000*l.* distributed as follows:

(i) 20,000*l.* of this is required to provide at first for 100 post-graduate research scholarships and research fellowships in science, technology, commerce, and economics for the best men from our universities and the higher technical institutions. These would be tenable for one year subject to renewal. It is suggested that the graduates should be nominated by the universities on a similar system to that which has been worked by the Carnegie Fund in Scotland with admirable results.

(ii) 40,000*l.* will be required to aid promising lines of research in engineering, electrical, chemical, and textile and other work where there are obviously many processes in which research may be promoted with great advantage, and invaluable indications have already been afforded as to promising lines of research. A series of grouped researches would be included in this branch of the scheme.

(iii) 30,000*l.* will be required to encourage the formation of specialised departments in association with existing universities or higher technological institutions or independently.

6. I have given special attention to the question of the administration of this part of my scheme, the object of which is to bring science to bear on industry and industry to make use of science. Having regard to the facts –

(1) That the great bulk of such research will be done in universities and colleges which are already aided by the Board;

(2) That the essence of research is cooperation between teachers and students; and

(3) That research is an integral part of an efficient system of higher education.

I have come to the conclusion that the Minister of Education should be responsible for this work. At the same time the cooperation of the principal trades and industries is essential to the success of this part of my proposals. I therefore recommend the formation of a Central Council of Commercial and Industrial Research, responsible to the Minister of Education. The Council should be as small as possible. It should not represent interests, but be composed of scientists, traders and other persons selected because of their personal fitness. This Council should have a large panel of advisers representing various industries, scientific and professional organisations, etc., selected members of which should be called upon for advice as required. In order to prevent overlapping and to promote cooperation, I suggest that the different Government Departments with research funds at their disposal should appoint representatives, who will be entitled as assessors to share in the deliberations of the Council. The Board of Trade should clearly be asked to participate. The work of the Central Council of Research would fall chiefly under these three heads:

(1) They would award the post-graduate and research scholar-ships which I propose, and would approve the subjects in which, and conditions under which, the researches should be undertaken, whether in laboratories, workshops, or other places.

(2) They would advise on and assist the individual or grouped researches mentioned in paragraph 5 (ii).

(3) They would also advise on and assist the provision and maintenance of specialised departments mentioned in para-graph 5 (iii).

7. I believe that a scheme on these lines, by bringing Science and Industry into immediate contact, would very soon create a demand for a larger supply of trained scientists for industry. To secure permanence and continuity in the carrying out of these purposes, I suggest that the Board of Education should have placed at its disposal by Act of Parliament the sum of 1,000,000*l.* for a period of five years, to be expended on the advice of the Council. A similar principle has been adopted in the case of the Development Fund, and is, I believe, essential to the success of this part of my scheme.

1.3.2 The State Meets the Scientists

> From Deputation from the Royal Society and the Chemical Society
> to meet the Presidents of the Board of Education and the Board of
> Trade, 9 May 1915, Board of Education, Public Record Office
> Ed. 24/1579.

> In May 1915 representatives from the Royal Society and the chem-
> ical societies presented separate Memorials to Joseph Pease
> (President of the Board of Education) and Walter Runciman
> (President of the Board of Trade) urging the government to provide
> increased funds to encourage a closer connexion between industry
> and academic scientists 'to watch the development of the chemical
> industry in the United Kingdom and advise manufacturers as to
> the most profitable lines to be followed up'. In response the
> scientists were informed of the government's plans to create a
> central advisory council for scientific and industrial research.

Professor E. Frankland, FRS: I think it would be necessary to go
back to the days of the late Prince Consort to find evidence and
symptoms of real understanding in any exalted personage of the vital
importance of chemistry to the real and sound welfare of the State.
Indeed, whatever position chemistry does occupy in the State
at the present time is for the most part, I think, a much
diminished inheritance of the recognition which it received by the
establishment under august patronage of the Royal College of
Chemistry in 1851; and I do not think that it is overstating the
case if I say that there is probably less interest in chemical science
manifested by the cultivated and ruling classes of this country
today than there was half a century ago, when the late Professor
Hoffmann presided over the Royal College of Chemistry, and
perhaps even less than there was one hundred years ago, when
Sir Humphrey Davy was drawing distinguished Society into the
premises of the Royal Institution in Albemarle Street. I cannot
call to mind, for instance, any single occasion upon which I have
ever seen a Member of the Government present at a chemical
discourse, however eminent the speaker, and however epoch-
making the subject of the communication. I do not wish to suggest,
of course, that it is to be expected that any Member of the Government must necessarily have any real interest in science, but I think
that the occasional presence of a Cabinet Minister at a scientific
gathering would have much the same sort of effect as that of the

duty which is imposed upon His Majesty the King to let himself be seen at Epsom, or at Ascot, or at Goodwood, in order to promote the interests of British horse-breeding. This profound neglect of science, and of chemical science in particular, is, I think, something specifically British, and I believe it would be difficult to find any other civilised country in which a comparable state of affairs exists. It is only since the beginning of the war that the general public has begun faintly to realise how dependent modern civilisation is on the creative forces of chemical science. It came as a surprise to many of the dwellers in what I regret to have to call the fools' paradise of England that a number of indispensable commodities were no longer obtainable because their supply was stopped from the country which, above all others, has consistently and systematically fostered the study of chemistry and its application. Can anything more humiliating be imagined than that the greatest Empire which the world has ever seen should have found that some of the commonest and most important drugs could no longer be dispensed, that the uniforms of our gallant soldiers could not be dyed to a constant colour, that our laboratories were crippled and paralysed by the want of some of the most important reagents and pieces of apparatus, that we were surrounded by the gravest difficulties in the way of obtaining the necessary materials for the manufacture of high explosives, and lastly I would venture to think that to the same category belongs the fact that you, in your place in Parliament, were obliged to admit that the Government had granted licences to trade with the enemy. An impartial investigation into the present position of the chemical industries reveals the fact that nearly all those industries which involve a very profound knowledge of chemical science, and the continuous employment of chemical research, have either left our shores or have never been established within this realm, and that those chemical industries which still remain, and some of which flourish, do so, however, not through any special knowledge or any special aptitude or any special merits that we possess, but because, for instance, we were first in the field, or because we enjoy advantages in the way of raw materials or geographical position, or some other factors of that kind. It is the earnest desire of the Chemical Society, which has sent us on this Deputation today, that if possible a halt should be called to this systematic destruction of our chemical industries which we have witnessed going on before us for two generations past, and of which some of the most eminent members of our profession have given warning times without number, but almost invariably without any effect at all.

REPLY

Mr Runciman: Gentlemen, my colleague and I have listened with care to what you have said this afternoon. I feel sure that he will agree with me when I say that you have much too humble an opinion of your own achievements and the distinguished position which you occupy ... I know of no great manufacturer at the present moment who has anything like a wide outlook on the possibilities of the future who is not as perturbed as you are, and as we are, at the shortage of young chemists who will take their place when our generation has passed away. I have had many opportunities during the last nine or ten months, as Dr Forster has remarked, of realising how short we are compared with Germany of men of great chemical ability and experience. The numbers that we have are really of no credit to us. There is at the top of the profession a considerable team of most able men, and I am glad to think that even the Badische works are prepared to accept the advice of an English chemist when they wish to deal with some of the more refined portions of their industry; but the number of distinguished men is small, and I am correctly interpreting the knowledge that you have, and which you have to some extent expressed this afternoon, when I say that it is in the second grade, and amongst the junior ranks, where we are most alarmingly short. We fully realise that, and we realise that the shortage is due to the fact that the career of a chemist in this country is most precarious. I have been astounded to come across manufacturers who were dependent for their success upon chemical knowledge still regarding, I will not say £120, but certainly £250 or £300, as ample remuneration for the man on whom their whole success in the future must depend. We must certainly alter all that. . . . What we are anxious to do now is to enlist the best and the most highly trained brains and the most experienced intellects in order to advise us in connexion with a scheme which we are considering as a Government. We as a Government feel that if we are to do anything we must have money. Directly we come to the distribution of anything like large sums, the House of Commons naturally desires to take an interest in the expenditure of that money and to hold a Minister responsible for that expenditure. The Government has already shown its appreciation of the work of some of the scientific bodies. The Royal Society, I think, receives £4,000 a year, and the National Physical Laboratory £7,000 a year, and amounts of that kind are distributed by the Government; but if we are to do anything real to place us in a proper position, we shall require very much larger sums than those in the future. Therefore I think the Government must be responsible for the expenditure of money, but it is most important that the Government should be able to have the advice of those great independent bodies which they do not desire to absorb in any way, which they do not desire to interfere with, but who can render them very great assistance in connexion with their duties. You have put before us a scheme in connexion with the chemists, and you have asked us to help to establish a central advisory committee – what you call an intelligence department . . . but what I can say to you is that we do realise the importance of research work; we have now secured the consent of the Government to proceed with a scheme, we are asking all the scientists in the country to help us, and to cooperate in one way and another, and such schemes as Dr Forster has placed before us will receive our very careful consideration. We do not want to act in a hurry, but, on the other hand, we do intend to act at once, and to try to establish machinery. We shall not have much money this year, but I hope as soon as the war is over very substantial sums may be forthcoming which will really enable us to take advantage of the machinery which we can establish at once. I am authorised to proceed with the scheme. I hope next week to be able to place it before my Parliamentary colleagues when the Board of Education Estimates will come under discussion, and in the meantime I must thank you for the way in which you have presented your case this afternoon, and to tell you that we are heartily with you in endeavouring to promote the objects which you have come here to further. (Applause.)

1.3.3 Alternative Proposals for the Government of Science

From Sir William Ramsay, 'Science and the State', *Nature*, 20 May 1915, vol. 95, pp. 309–11.

Sir William Ramsay (1852–1916), Professor of general chemistry at University College London from 1887 to 1913 and a Nobel prizewinner for his research on the noble gases, was one of the leading statesmen of science. Ramsay believed that insufficient provision had been made in the government's proposed arrangements for the dialogue between chemists, industrialists and government. His unique proposals for a Chemical Council of State reflected this belief. In correspondence with Joseph Pease and Arthur Henderson (Pease's successor as President of the Board of Education) Ramsay urged the government not to be afraid to 'trust the expert'.

On the evening of the same day on which the report [of the deputation to see Runciman and Pease (see 1.3.2)] appeared in *Nature*, Mr Pease announced in the House the intention of the Government to create an Advisory Council on Industrial Research – a committee of experts who would be able to consult with other expert committees working in different directions, and associated with leaders of industry. 'He was now considering names.'

Now, we do not doubt the good will of the members of the Cabinet, but we distrust their judgement in this matter. The handling of the dye scheme was, to say the least of it, very unfortunate. There are two German works, one the Mersey works, a branch of the 'Badische', and one at Ellesmere Port, a branch of the German works at Höchst, which might have been associated forcibly with this combine with advantage to the country; moreover, the total lack of chemical talent on the directorate does not argue for its success, as the public has testified by failing to subscribe the issue. The two 'eminent chemists' who advised the Government on the dye scheme doubtless do not thank the member of the Cabinet for their unsought publicity. These and other similar instances lead us to mistrust the judgement of Mr Pease and his colleagues on questions involving science.

There is certainly room for a chemical council, and I had already prepared a draft scheme about two months ago, which has been submitted to, and has had the general approval of, several of our leading industrial chemists. Perhaps it might help were an outline of the scheme to be given here. It is headed 'A Draft Scheme of a Chemical Council of State'. The clauses are as follow:

1. The dependence of the welfare of a country on its chemists

is obvious. Chemistry lies at the basis of practically all manufactures. Continental nations and the United States have long acknowledged this.

2. Great Britain is behind no nation in the eminence of its chemists. But inducements are lacking to persuade young men to accept minor positions. Chemical research, as indeed all research, is of two kinds; capital discoveries are made by some, and in this Great Britain probably leads. But the patient development of known ground requires men of a different calibre – men of more ordinary attainments; such men are lacking in numbers in Great Britain.

3. Men of the first rank exist both in the universities and in industry. It is of these that the Chemical Council should consist. About twenty men of this class could easily be named, of the highest reputation and of great experience. Among them every one of the numerous branches of chemistry could be covered; one or more of them would be competent to give expert opinion on every subject which falls within the purview of chemistry.

4. The manufacturing chemists in Great Britain are, generally speaking, not combined. It is true that the alkali-makers work on a mutual understanding; so do the ironmasters. But chemical products are so varied that it may be truly said that industrial chemists work in isolation from each other. It is also generally true that there is little contact between industrial and scientific chemistry. The teachers and students in universities and colleges know little of what passes in the world of manufacture, nor do industrial chemists, as a rule, consult the heads of scientific laboratories. This, again, does not obtain abroad.

5. A Chemical Council for the United Kingdom or for the Empire should comprise both classes of men: scientific investigators and those who apply scientific discoveries to industry. It should contain about twenty-four members, of whom one-third should be technical chemists, one-third scientific investigators, and one-third analytical and consulting chemists.

6. Its duties should be:

(a) To ascertain from every chemical factory in the kingdom (1) the nature of its raw material; (2) the nature and amount of its finished products; (3) the nature of its by-products and what becomes of them. Also to learn by inquiry of the purchasers and users of chemical products – (1) what articles they obtain from home manufacturers; (2) what articles they purchase from abroad; and (3) the causes which induce them to encourage foreign rather than home industries.

(*b*) To establish connexion with the chemical laboratories of universities and colleges, and to bring chemical researchers into contact with manufacturers, so that the latter should indicate to the former what problems await solution; and the former should keep the latter posted in any discoveries which appear to promise to be of technical value.

(*c*) To advise the Government on questions involving a skilled knowledge of chemistry and its applications.

10. The Committee should report once a year at least, or even at shorter intervals, to the Crown. It would appear advisable not to attach it to any Government department, but to associate it with the Board of Trade, the Board of Agriculture and Fisheries, the Local Government Board, the Board of Education, and also with the Government laboratories.

11. As it is clearly of advantage that such a committee should be non-political, it would be well if it were appointed by and were directly responsible to the Crown.

To whom is the nomination of the first members of such a committee to be entrusted? For on that will depend its success or its failure. I suggest that the President of the Royal Society, himself a most distinguished chemist, should be asked to nominate from the Fellows four persons, two scientific chemists and two technical chemists; and that they, under his chairmanship, should select the names of twenty other persons, themselves constituting four members of the Council. It is unlikely that Sir William Crookes could be prevailed on to add to his numerous onerous duties by himself serving on the Council; but he would probably consent to act as chairman of the electoral committee.

It is earnestly to be hoped that members of the Government will agree to adopt some such scheme. To embark without expert – real expert – advice on nominating the members of such a Council would be to expose it to risks equal to that attending the dye scheme, and would make it impossible to achieve the objects which they appear to have at heart. Let us hope that they will, in this case at least, trust the expert.

1.3.4 Two Views on the Neglect of Science in the War Effort

From H. G. Wells and Professor J. A. Fleming, letters to *The Times*, 11 and 15 June 1915.

By June 1915 informed public opinion had become aware of the disastrous implications of the apparent neglect of scientific expertise by the government in the conduct of the war. H. G. Wells precipitated an outcry in *The Times* with a letter condemning the government's lack of imagination in conserving scientific manpower. A few days later, J. A. Fleming, FRS, Professor of Electrical Engineering at University College, London, complained bitterly of the government's failure to mobilise the talent of scientists.

Sir,—We have reconstructed our Government and it is not for an innocent Englishman outside the world of politicians to estimate the advantages and disadvantages of the rearrangement of the House of Commons. But there is a matter beyond the range of party politics which does still seem to need attention and which has been extraordinarily disregarded in all the discussion that has led to the present Coalition, and that is the very small part we are still giving the scientific man and the small respect we are showing scientific method in the conduct of this war. I submit that there is urgent need to bring imaginative enterprise and our utmost resources of scientific knowledge to the assistance of the new-born energies of the Coalition; that this is not being done and that until it is done this war is likely to drag on and be infinitely more costly and infinitely less conclusive than it could and should be.

Modern war is essentially a struggle of gear and invention. It is not war under permanent conditions. In that respect it differs completely from pre-Napoleonic wars. Each side must be perpetually producing new devices, surprising and outwitting its opponent. Since this war began the German methods of fighting have been changed again and again. They have produced novelty after novelty, and each novelty has more or less saved their men and unexpectedly destroyed ours. On our side we have so far produced hardly any novelty at all, except in the field of recruiting posters. It is high time that our rulers and our people came to recognize that the mere accumulation of great masses of young men in khaki is a mere preliminary to the prosecution of this war. These masses make the body of an army, but neither its neck, head, nor hands, nor feet. In the field of aviation, for which the English and French temperaments are far better adapted than the German, there has been no energy of organisation at all. There has been great individual gallantry and a magnificent use of the sparse material available, but no great development. We have produced an insufficient number of aviators and dribbled out an inadequate supply of machines. Insufficient and inadequate, that is to say, in relation to such a war as this. We have taken no steps to produce a larger and more powerful aeroplane capable of overtaking, fighting, and destroying a Zeppelin, and we are as far as ever from making any systematic attacks in force through the air. Our utmost achievements have been made by flights of a dozen or so machines. In the matter of artillery the want of intellectual and imaginative enterprise in our directors has prevented our keeping pace with the German improvements in trench construction; our shortness of high explosives has been notorious, and it has led to the sacrifice of thousands of lives. Our Dardanelles exploit has been throughout unforeseeing and uninventive; we have produced no counterstroke to the enemy's submarine, and no efficient protection against his improved torpedoes. We have still to make an efficient use of poison gas and of armoured protection in advances against machine-guns in trench warfare. And so throughout almost the entire range of our belligerent activities we are to this day being conservative, imitative, and amateurish when victory can fall only to the most vigorous employment of the best scientific knowledge of all conceivable needs and material.

One instance of many will serve to show what I am driving at. Since this war began we have been piling up infantry recruits by the million and making strenuous efforts to equip them with rifles. In the meantime the actual experiences of the war have been fully verifying the speculations of imaginative theorists, and the Germans have been learning the lesson of their experiences. The idea that for defensive purposes one well-protected skilled man with a small machine-gun is better than a row of riflemen is a very obvious one indeed, but we have disregarded it. The Germans are giving up the crowding of men for defence purposes (though the weakness of the national quality obliges them still to mass for attacks), and they are entrusting their very small and light machine-guns in many cases to officers. They have, in fact, adopted as their 1915 model of trench defence the proper scientific thing. Against this we fire our shrapnel and hurl our infantry.

Now these inadequacies are not incurable failures. But they are likely to go on until we create some supplementary directive force, some council in which the creative factors in our national life, and particularly our scientific men and our younger scientific soldiers and sailors, have a fuller representation and a stronger influence than they have in our present Government. It is not the sort of work for which a great legal and political career fits a man. That training and experience, valuable as it is in the management of men and peoples, does indeed very largely unfit men for this incessantly inventive work. A great politician has no more special aptitude for making modern war than he has for diagnosing diseases or planning an electric railway system. It is a technical business. We want an acting sub-Government of scientific and technically competent men for this highly specialised task.

Such a sub-Government does in effect exist in Germany. It is more and more manifest that we are fighting no longer against that rhetorical system of ancient pretensions of which the Kaiser is the figure-head. In Flanders we are now up against the real strength of Germany; we are up against Westphalia and Frau Krupp's young men. Britain and France have to get their own brilliant young engineers and chemists to work against that splendid organisation. Unless our politicians can add to the many debts we owe them, the crowning service of organising science in war more thoroughly than they ever troubled to do it in peace, I do not see any very great hope of a really glorious and satisfactory triumph for us in this monstrous struggle.

Very sincerely yours,

H. G. Wells.

Sir,—It would be difficult to overstate the importance of the discussion on the above subject in your columns which was initiated by the letter of Mr Wells. At the present moment the scientific ability of this country is in the position of the magnetic molecules of iron as regards external magnetic effect. They produce no result until some external magnetic force compels them all to orient themselves in the same direction.

There is no want of ability, but there is an entire absence of external directing power. Nay, rather, special steps have been taken to inhibit scientific activity in directions which might assist the nation.

Take, for instance, the subject of electric waves, which might be used as an implement of warfare in certain ways I forbear to point out for obvious reasons. It was unquestionably right of the General Post Office to put a stop at the outset of the war to all amateur wireless telegraphy, to prevent German spies conducting their communications with antenna wires put up a chimney.

Is it, however, an advantage to the country that all the expert knowledge on this subject outside certain official circles should be cast on one side and neglected?

A few days ago an eminent electrical engineer was sitting in my room here, and said to me, 'I am too old to enlist or even to do manual work in the manufacture of shells, but I have a considerable scientific knowledge which I am just yearning to employ in the service of the country, yet I cannot find any person in authority who will tell me how to do it.'

This sentence expressed concisely not only my friend's feelings but my own, and I am confident that of hundreds of other scientific men as well. At the present moment, after ten months of scientific warfare, I myself, although a member of several scientific and technical societies and a Fellow of the Royal Society as well, have not received one word of request to serve on any committee, cooperate in any experimental work, or place expert knowledge, which it has been the work of a lifetime to obtain, at the disposal of the forces of the Crown. It is not enough to make vague suggestions as to the detection of submarines or destruction of Zeppelins. Rough ideas have to be hacked into shape, reduced to practice, and tested on a large scale. All this means organisation, expenditure, assistants, and definite practical experiments. It seems to demand a special Government department, which shall enlist in its service trained and experienced investigators for definite ends. This war will be won in the laboratories and workshops almost as much as in the field, and it will only be won when the Government organise the scientific intellect of the country as well as the manual labour with that single purpose.

I am, &c.,

J. A. Fleming.

1.3.5 The Reconciliation of Science, Research and Industry

From Board of Education, Scheme for the Organisation and Development of Scientific and Industrial Research, Cd. 8005, 25 July 1915.

The work of the new organisation was outlined in the White Paper. Knowledge that the government had set up the Advisory Council to accelerate the application of research to industry and to provide opportunities for more research to be done in the universities did much to calm the fears voiced by Wells and Fleming [1.3.4].

1. There is a strong consensus of opinion among persons engaged both in science and in industry that a special need exists at the present time for new machinery and for additional State assistance in order to promote and organise scientific research with a view especially to its application to trade and industry. It is well known that many of our industries have since the outbreak of war suffered through our inability to produce at home certain articles and materials required in trade processes, the manufacture of which has become localised abroad, and particularly in Germany, because science has there been more thoroughly and effectively applied to the solution of scientific problems bearing on trade and industry and to the elaboration of economical and improved processes of manufacture. It is impossible to contemplate without considerable apprehension the situation which will arise at the end of the war unless our scientific resources have previously been enlarged and organised to meet it. It appears incontrovertible that if we are to advance or even maintain our industrial position we must as a nation aim at such a development of scientific and industrial research as will place us in a position to expand and strengthen our industries and to compete successfully with the most highly organised of our rivals. The difficulties of advancing on these lines during the war are obvious and are not under-estimated, but we cannot hope to improvise an effective system at the moment when hostilities cease, and unless during the present period we are able to make a substantial advance we shall certainly be unable to do what is necessary in the equally difficult period of reconstruction which will follow the war.

2. The present scheme is designed to establish a permanent organisation for the promotion of industrial and scientific research.

It is in no way intended that it should replace or interfere with the arrangements which have been or may be made by the War Office or Admiralty or Ministry of Munitions to obtain scientific advice and investigation in connexion with the provision of munitions of war. It is, of course, obvious that at the present moment it is essential that the War Office, the Admiralty, and the Ministry of Munitions should continue to make their own direct arrangements with scientific men and institutions with the least possible delay.

3. It is clearly desirable that the scheme should operate over the Kingdom as a whole with as little regard as possible to the Tweed and the Irish Channel. The research done should be for the Kingdom as a whole, and there should be complete liberty to utilise the most effective institutions and investigators available, irrespective of their location in England, Wales, Scotland or Ireland. There must therefore be a single fund for the assistance of research, under a single responsible Body.

4. The scheme accordingly provides for the establishment of:

(a) A Committee of the Privy Council responsible for the expenditure of any new moneys provided by Parliament for scientific and industrial research;

(b) A small Advisory Council responsible to the Committee of Council and composed mainly of eminent scientific men and men actually engaged in industries dependent upon scientific research.

5. The Committee of Council will consist of the Lord President, the Chancellor of the Exchequer, the Secretary for Scotland, the President of the Board of Trade, the President of the Board of Education (who will be Vice-President of the Committee), the Chief Secretary for Ireland, together with such other Ministers and individual Members of the Council as it may be thought desirable to add.

The first non-official Members of the Committee will be:

The Right Hon. Viscount Haldane of Cloan, OM, KT, FRS,

The Right Hon. Arthur H. D. Acland, and

The Right Hon. Joseph A. Pease, MP.

The President of the Board of Education will answer in the House of Commons for the sub-head on the Vote, which will be accounted for by the Treasury under Class IV, Vote 7, 'Scientific Investigations, &c.'.

It is obvious that the organisation and development of research is a matter which greatly affects the public educational systems of the Kingdom. A great part of all research will necessarily be done in Universities and Colleges which are already aided by the State, and the supply and training of a sufficient number of young persons competent to undertake research can only be secured through the public system of education.

6. The primary functions of the Advisory Council will be to advise the Committee of Council on –

(i) proposals for instituting specific researches;
(ii) proposals for establishing or developing special institutions or departments of existing institutions for the scientific study of problems affecting particular industries and trades;
(iii) the establishment and award of Research Studentships and Fellowships.

The Advisory Council will also be available, if requested, to advise the several Education Departments as to the steps which should be taken for increasing the supply of workers competent to undertake scientific research.

Arrangements will be made by which the Council will keep in close touch with all Government Departments concerned with or interested in scientific research and by which the Council will have regard to the research work which is being done or may be done by the National Physical Laboratory.

7. It is essential that the Advisory Council should act in intimate cooperation with the Royal Society and the existing scientific or professional associations, societies and institutes, as well as with the Universities, Technical Institutions and other Institutions in which research is or can be efficiently conducted.

It is proposed to ask the Royal Society and the principal scientific and professional associations, societies and institutes to undertake the function of initiating proposals for the consideration of the Advisory Council, and a regular procedure for inviting and collecting proposals will be established. The Advisory Council will also be at liberty to receive proposals from individuals and themselves to initiate proposals.

All possible means will be used to enlist the interest and

secure the cooperation of persons directly engaged in trade and industry.

11. The Advisory Council will proceed to frame a scheme or programme for their own guidance in recommending proposals for research and for the guidance of the Committee of Council in allocating such State funds as may be available. This scheme will naturally be designed to operate over some years in advance, and in framing it the Council must necessarily have due regard to the relative urgency of the problems requiring solution, the supply of trained researchers available for particular pieces of research, and the material facilities in the form of laboratories and equipment which are available or can be provided for specific researches. Such a scheme will naturally be elastic and will require modification from year to year; but it is obviously undesirable that the Council should live 'from hand to mouth' or work on the principle of 'first come first served', and the recommendations (which for the purpose of estimating they will have to make annually to the Committee of Council) should represent progressive instalments of a considered programme and policy. A large part of their work will be that of examining, selecting, combining, and coordinating rather than that of originating. One of their chief functions will be the prevention of overlapping between institutions or individuals engaged in research. They will, on the other hand, be at liberty to initiate proposals and to institute inquiries preliminary to preparing or eliciting proposals for useful research, and in this way they may help to concentrate on problems requiring solution the interest of all persons concerned in the development of all branches of scientific industry.

1.3.6 Science in National Affairs

From *Nature*, 21 October 1915, vol. 96, p. 195.

In October 1915, a few months after the Advisory Council for Scientific and Industrial Research had been set up, *Nature* attacked its failure to fulfil its early promise. In particular the journal attacked the administrators' amateurism and the lack of scientific professionals on the Council's staff.

Under a democratic constitution it is perhaps too much to expect that Parliament will pay much attention to scientific men or methods; yet, as was shown in the debate upon the scheme for the constitution of an advisory council of scientific and industrial research last May, the members of the House of Commons are ready to support plans for bringing science in closer connexion with industry. The monies provided by Parliament for this purpose are to be under the control of a committee of the Privy Council, which will be advised by a council constituted of scientific and industrial experts. The scheme was conceived rightly enough, but when it passed into the hands of officials of the Board of Education much of its early promise was lost. Most people would regard it as essential that the executive officers of a council concerned with the promotion of industrial research should know what is done in this direction in other countries, and have sufficient knowledge of science and industry to formulate profitable schemes of work. The success of such a body depends largely upon the initiative of the secretary; and in an active and effective council we should expect him to be selected because of close acquaintance with problems of industrial development along scientific lines. But what is the position in this case? The scheme is issued by the President of the Board of Education, Mr Arthur Henderson, a Labour member, who owes his post entirely to political exigencies, the secretary to the committee of the Privy Council is the Secretary of the Board, Sir Amherst Selby-Bigge, whose amiability is above reproach, but who knows no more of practical science and technology than a schoolboy, and the secretary of the Advisory Council is Dr H. F. Heath, whose interests are similarly in other fields than those of science.

The belief that the expert – whether scientific or industrial – has to be controlled or guided by permanent officials having no special knowledge of the particular subject in hand is typical of our executive system. While such a state of things exists, most of the advantages of enlisting men of science for national services must remain unfulfilled. The various scientific committees which have been appointed recently have, we believe, been able to give valuable aid in connexion with problems submitted to them, but they would be far more effective if the chiefs of the departments with which they are associated possessed a practical knowledge of scientific work and methods. Without such experience the executive is at the mercy of every assertive paradoxer and cannot discriminate between impracticable devices and the judgement of science upon them. While, therefore, the country has at its disposal the work – either voluntary or nearly so – of experts in all branches of applied science, it cannot use these services to the best advantage unless the departments concerned with them have scientific men among the permanent officials; and that is not the case at present.

1.3.7 The Advisory Council Plans for the Future

From the Report of the Committee of the Privy Council for Scientific and Industrial Research for 1915–16, Cd. 8336, 1916, pp. 9, 10–11.

In their first report the Advisory Council outlined their plan of campaign. The immediate problem was to convince the industrialists of the importance of fundamental research; the second problem was to improve communication between science and industry; the third was to provide more opportunities for research in the universities.

The necessity for the central control of our machinery for war had been obvious for centuries, but the essential unity of the knowledge which supports both the military and industrial efforts of the country was not generally understood until the present war revealed it in so many directions as to bring it home to all. War has remained as much an art as ever, but its instruments, originally the work of the craftsman and the artist, are now not only forged by the man of science; they need a scientific training for their effective use. This is equally true of the weapons of industry. The brains, even the very processes, that today are necessary to the output of munitions were yesterday needed, and will be needed again tomorrow, for the arts of peace.

This is the central fact which justifies the establishment of the new machinery in the midst of a struggle that is absorbing the whole energies of the nation in a way no previous war has done. . . .

The Method of Approach

Accordingly the Council decided to give science in its applications to industry precedence over pure science in their deliberations. They are under no misapprehension as to the relations between pure and applied science. It has been said that what people call applied science is nothing but the application of pure science to particular classes of problems. And, properly speaking, this no doubt is so; there are not two different kinds of science. At the same time the Council realise that they have to deal with the practical business world, in whose eyes a real distinction seems to exist between pure and applied science. The average manufacturer is impressed with the importance of quick returns; he cannot afford to wait. The managing director of one manufacturing firm recently told us that he had no interest in research which did not produce results within a year. If science can help him to overcome the difficulties that cross his path from day to day he welcomes her. He wants a handy servant, not a partner with ideas of her own. This is the position from which a start must be made. No doubt there are firms much more enlightened than this – great enterprises which spend lavishly on fundamental investigations which will, if successful, not merely remove a hitch, or improve output, but reveal the exact nature of a process and give the manufacturer assured control of his material. Yet there is a natural and obvious reason for the scepticism of the average businessman. Apart from the long period of work and preparation which a properly organised research laboratory attached to a factory involves before important results can be looked for, and apart from the risk that an unsuitable director of such a laboratory may involve a firm in heavy pecuniary loss, there is the plain fact that 'the difficulties that present themselves to manufacturers or merchants seldom afford an indication of the true nature of the problems to be solved. They are generally secondary in their nature, and a direct attack on them is likely to be as empirical as the symptomatic treatment of disease.'

Thus such quick-result inquiries as the manufacturer is induced to make are very likely to be fruitless and to reduce his enthusiasm for science still further. She finds too few suitors in our industries for she is a mistress who reserves her favours for a complete and single-hearted devotion.

But there was another consideration which alone would have compelled the Council to begin with research of directly industrial application. The Universities, which are the natural homes of research in pure science, have been so depleted of both students and teachers by the war that they are barely able to continue their routine work and can command at the moment neither the leisure nor the detachment of spirit that are essential conditions of original research. Inquiries we have recently made in the hope that we might be able to help individual investigators at the Universities to increase their output, have convinced us that any effective encouragement of research in pure science must await the return of peace.

How then, were the Council to proceed? On the one hand the war had greatly reduced the number of workers available for research of any kind. On the other, every consideration of urgency pointed to concentration on the field of industrial research both as the best means of enlisting the cooperation of the manufacturers, without which no Government scheme could succeed, and as the quickest way of preparing for the trade conditions likely to arise after the war. In these circumstances the Council might have concentrated all their energies upon the rapid initiation of one or two far-reaching proposals likely to attract wide attention and designed to improve the outlook in one or other of our greatest industries. But the problems with which we have to deal are not susceptible of rapid and dramatic solution. The successful organisation of research on a national scale is likely to take longer than the establishment of a good works laboratory – a long business at best. The only hopeful line of action appeared therefore to involve a period of careful inquiry and consultation, followed by a gradual and systematic attack upon a wide and carefully selected front.

> This 'systematic attack', involved five separate activities: (a) aid to existing researches being conducted by professional associations; (b) conferences with professional and other societies; (c) a register of researches being conducted at the outbreak of war; (d) aid to research in educational institutions in the form of grants to post-graduate students; and (e) the formation of standing committees to survey certain fields of research, such as metallurgy and engineering.

1.3.8 The Origins of the 'Million Fund' and the Department of Scientific and Industrial Research.

From the Report of the Committee of the Privy Council for Scientific and Industrial Research for 1916–17, Cd. 8718, 1917–18.

In October 1916, just over one year after the formation of the Advisory Council for Scientific and Industrial Research, the government agreed to make £1 million available for the promotion of cooperative research in industry. The fund was to be entirely separate from the budget of the Advisory Council. Bearing in mind the growing responsibilities of the Council and also the additional financial responsibility of managing the 'Million Fund' it was decided to create a separate Department for Scientific and Industrial Research, responsible to the Lord President of the Council.

2. On 1 December last Lord Crewe, the Lord President of the Council, announced to a deputation from the Board of Scientific Societies that the work of the Committee of Council for Scientific and Industrial Research with that of their Advisory Council, first instituted for reasons of convenience under the aegis of the Board of Education, had been established as a separate department with offices in Great George Street. As the work of the Advisory Council developed and the industrial side of research grew in bulk and importance, it became clear that a separate organisation having its own estimates in charge of a minister responsible to Parliament was a necessity. The foundation of the new department led to the creation under Your Majesty's Sign Manual of the Imperial Trust for the Encouragement of Scientific and Industrial Research. The Charter empowers the Lord President and six other high officers of State to hold funds, to enter into contracts and agreements and to do other things in furtherance of the objects of the Committee of Council for Scientific and Industrial Research. A donation of £1,000 made by Messrs R. H. and R. Williamson to be expended as to half its amount on research into some subject of mechanical engineering in conjunction with the Institution of Mechanical Engineers and as to the other half on such research as the Committee of Council may determine, has been deposited with the Trust as well as a sum of £13 14s. 9d., representing the turnover on the Treasury Bill, which was drawn for the donation.

3. It is also intended that the Trust shall hold on behalf of the Department the sum of one million sterling which Parliament has voted for the purposes of the Department. The negotiations of our Advisory Council with the leading manufacturers in the various industries have shown that it would not be possible to develop systematic research on a large scale unless the Government were in the position to assist financially over an agreed period of years. The industries and especially the great staple industries might be expected to bear a considerable share of the large sums involved; but

on the one hand it would be difficult if not impossible to foresee from year to year the amount of expenditure likely to be called for and on the other the industries would not unreasonably look for an assurance that, as the need arose, the Department would be in the position to give the necessary aid. Further we were advised that at any rate the larger and more prosperous industries might be expected, after an initial impetus, to find themselves both willing and able to continue the work of research without direct assistance from the State. These three considerations convinced the Government that the somewhat novel expedient of placing a fund at our disposal to be spent over a period of five or six years afforded the best means of dealing with the problem. This fund, amounting to a million pounds, will be deposited with the Imperial Trust and expended in accordance with our directions. Our Advisory Council have recommended that the money thus made available should be spent in the form of grants in aid of research undertaken by firms in any industry which may combine to conduct it on a cooperative basis. The Council advise us that the best means to this end is the establishment under the Companies' Acts of Associations for Research limited by guarantee and trading without profit. We have approved this method of procedure which has the additional advantage that the Board of Inland Revenue have decided, with the approval of the Chancellor of the Exchequer, that no objection shall be offered by their Surveyor of Taxes to the admission, as a working expense for Income Tax allowance, of contributions by Traders to Industrial Associations under Government supervision which may be formed for the sole purpose of scientific research for the benefit of the various trades; and that the allowance would be equally applicable to traders' contributions specifically earmarked for the sole purpose of the Research Section of an adapted existing Association.

1.4 SCIENTIFIC EDUCATION

1.4.1 Science in Education: The Civil Service Examinations

From *The Neglect of Science*. Proceedings of a Conference held on 3 May 1916 (Harrison & Sons, London, 1916).

A memorandum, published in *The Times* on 2 February 1916, drew public attention to the neglect of science in government administration. Among the distinguished signatories were Sir William Crookes (former President of the Royal Society), Lord Rayleigh, Professor Edward Perkin and Sir Edward Thorpe. The memorandum concluded, strongly, that the vital link in the establishment of a scientifically aware Civil Service was the Civil Service examinations, hitherto confined to traditional, non-scientific subjects. The memorandum urged that scientific subjects be awarded an equal share of marks in the competitive examinations. The memorandum was sent to scientists, educationalists and industrialists and formed the basis of a conference held in May of the same year in London, at which Lord Rayleigh took the chair.

'The Neglect of Science'

With a full sense of responsibility we submit the following memorandum on a subject which, we are convinced, requires immediate attention and drastic action. It concerns the public interests, and the public alone can deal with it.

It is admitted on all sides that we have suffered checks since the War began, due directly as well as indirectly to a lack of knowledge on the part of our legislators and administrative officials of what is called 'science' or 'physical science'. By these terms we mean the ascertained facts and principles of mechanics, chemistry, physics, biology, geography, and geology. Not only are our highest Ministers of State ignorant of science, but the same defect runs through almost all the public departments of the Civil Service. It is nearly universal in the House of Commons, and is shared by the general public, including a large proportion of those engaged in industrial and commercial enterprise. An important exception to this rule is furnished by the Navy and also by the medical service of the Army; in both these services success has been achieved by men who, while in no way inferior in courage, devotion, and self-sacrifice to their brethren elsewhere, have received a scientific training.

This grave defect in our national organisation is no new thing. It has been constantly pressed upon public attention during the last fifty years as a cause of danger and weakness. In the whole history of British Governments there has been only one Cabinet Minister who was a trained professional man of science – the late Lord Playfair.

It is not our intention here to enumerate the catalogue of specific instances in which a want of understanding of 'physical science' has led the Ministry and Executive into error. This has been done elsewhere, but as an example of the ignorance which we deplore we may instance the public statement of a member of the Government – unchallenged when made – that his colleagues should be excused for not having prevented the exportation of lard to Germany, since it had only recently been discovered that glycerine (used in the manufacture of explosives) could be obtained from lard. The fact is, on the contrary, that the chemistry of soap-making and the accompanying production of glycerine is very ancient history. In order that such serious blunders may be avoided it is essential that we should have a proportion of men in the Government who, if not actual experts, yet have such a knowledge of science as will give them an intelligent respect for it, and an understanding of what it can do, how to make use of it, and to whom to apply when special knowledge is required.

Our success now, and in the difficult time of reorganisation after the War, depends largely on the possession by our leaders and administrators of scientific method and the scientific habit of mind. They must have knowledge and the habit of promptly applying known means to known ends. To trust to luck is a mark of the dangerous complacency bred of ignorance. The evidence of those back from the Front makes it clear that, as of old, our 'people are destroyed for lack of knowledge'.

How can such a revolution as we desire in the higher and lower grades of the public service be brought about? Obviously it can only be effected by a great change in the education which is administered to the class from which these officials are drawn. The education of the democracy, which gives its consent to the present state of things, would follow the change in the education of the wealthier classes. For more than fifty years efforts have been made by those who are convinced of the value of training in experimental science to obtain its introduction into the schools and colleges of the country as an essential part of the education given therein. At first it seemed as though the effort had been successful, but it is clear that the old methods and old vested interests have retained their dominance, at least as far as our ancient Universities and great schools are concerned. At Cambridge but four colleges are presided over by men of scientific training; at Oxford not one. Of the thirty-five largest and best-known public schools thirty-four have classical men as head masters. Science holds no place in the list.

The examinations for entrance into Oxford and Cambridge, and for appointments into the Civil Service and the Army, are among the greatest determining factors in settling the kind of education given at our public schools. Natural science has been introduced as an optional subject for the Civil Service examinations, but matters are so arranged that only one-fourth of the candidates offer themselves for examination in science. It does not pay them to do so; for in Latin and Greek alone (including ancient history) they can obtain 3,200 marks, while for science the maximum is 2,400, and to obtain this total a candidate must take four distinct branches of science. For entrance into Woolwich, science has within the last few years been made compulsory, but for Sandhurst it still remains optional. This college is probably the only military institution in Europe where science is not included in the curriculum. The result of this system of examination not merely upon the successful candidates, but upon all the great schools and the old Universities, which necessarily (as things are at present arranged) work with them in aim and interest, is a neglect of the study of the natural sciences, and to some extent an indifferent, not to say contemptuous, attitude towards them.

The one and effective way of changing this attitude and of giving us both better-educated Civil Servants and a true and reasonable appreciation of science in all classes is in the hands of the Legislature, and of it alone. If a Bill were passed directing the Civil Service Commissioners and Army Examination Board to give a preponderating – or at least an equal – share of marks in the competitive examination to natural science subjects, with safeguards so as to make them tests of genuine scientific education and not an incentive to mere 'cram', the object we have in view would be obtained. Science would rise in our schools to a proper position, and gain the respect necessary for the national welfare. A popular appreciation and understanding of science would begin to develop, and our officials of all kinds, no less than Members of Parliament, would come to be as much ashamed of ignorance of the commonplaces of science as they would now be if found guilty of bad spelling and arithmetic. Not at once, but little by little, the professional workers in science would increase in number and gain in public esteem. Eventually the Board of Trade would be replaced by a Ministry of Science, Commerce and Industry, in full touch with the scientific knowledge of the moment. Public opinion would compel the inclusion of great scientific discoverers and inventors as a matter of course in the Privy Council, and their occupation in the service of the State.

With this object in view we urge the electorates to insist that candidates for their suffrages should pledge themselves to aid by legislation in bringing about a drastic reform in the scheme of examinations for all the public services in the sense we have indicated.

Our desire is to draw attention to this matter, not in the interests of existing professional men of science, but as a reform which is vital to the continued existence of this country as a Great Power.

1.4.2 Scientific Education

From the Report of the Committee appointed to inquire into the Position of Natural Science in the Educational System of Great Britain, Cd. 9011, 1918.

The Committee under the chairmanship of Sir J. J. Thomson (1856–1940), Cavendish Professor of Physics at Cambridge, had been set up in 1916 to make a detailed study of provision for the teaching of natural science in British schools. In the following extract from the Introduction to the report the place of science in education is briefly reviewed historically.

2. Not for the first time our educational conscience has been stung by the thought that we are as a nation neglecting Science. Attention was called to this neglect by the Report of the Royal Commission on the nine Public Schools in 1864, when it was recommended that all boys should receive instruction in some branch of Natural Science during part at least of their school career. A Committee of the British Association dealt with the subject again in 1866, drawing the valuable distinction between scientific information and scientific training, and making recommendations which influenced the course of science teaching in schools. That there was need for these exhortations can be proved without any elaborate survey of the history of science teaching in England. In 1863, at the very time the Public Schools Commission was holding its enquiry, the only instruction in Science at one of the greatest schools in England was given on Saturday afternoon

by a visiting teacher, and his meagre apparatus was stored in so damp a cupboard that his experiments usually broke down. It is not surprising that the headmaster of this school told the Commissioners that 'instruction in physical sciences was, except for those who have a taste, and intended to pursue them as amateurs or professionally, practically worthless.' Steps had been taken before these dates at certain schools to introduce the teaching of Science, but this work was done under great difficulties and was regarded with jealousy by the staffs, with contempt by the boys and with indifference by the parents.

Gradually, thanks to these reports and to the efforts of gifted teachers within the schools, this Benjamin of subjects won toleration if not affection in the family circle. Meantime public interest in Science was being aroused by the achievements of scientific workers like Darwin and Kelvin and by the writings of Spencer, Kingsley, Tyndall and Huxley, and this interest was reflected in the schools.

During these years however secondary education was within the reach of but few. The big Public Boarding Schools – then to be numbered on the fingers of two hands – educated a limited number for whom a road had been made by family traditions or increasing wealth; the old established Grammar Schools scattered sparsely over the country offered to others in their immediate neighbourhood opportunities of education often most eagerly seized and fruitfully used; but boys, even though they found in most schools science teaching available if they sought it out, were sometimes denied it altogether, and they were certainly discouraged from pursuing it unless they had shown incapacity for Classics or Mathematics. For girls even these limited opportunities did not exist. Information about their education at this period is scanty, but it may safely be said that no organised instruction in Science was available for them. These weaknesses, which persisted long after the battle of Science was half won, have never been entirely removed by a great stirring of public opinion, even though our defects in scientific education have been fitfully pointed out and to some extent corrected.

Further, while the secondary school, so far as it existed, remained under the classical tradition, the schools which grew up under the Science and Art Department tended to be one-sided in the opposite direction, fostering Science to the exclusion of literature. The river of educational enthusiasm, never too strong, was consequently split into two weak streams.

The problem has, of course, been affected by the wide extension of secondary education that has marked the last fifteen years, but the older schools have not yet been entirely freed from all their prejudices, and the newer schools, in spite of their better balance of subjects, may perhaps have missed some of their opportunities.

3. From schools so few in number and so limited in aim recruits for the Universities could not be obtained in abundance. There were professors of scientific subjects at both Oxford and Cambridge all through the eighteenth and nineteenth centuries, and no doubt they attracted to their lecture rooms individual students, but it was not until half-way through the latter century that the establishment of Honour Schools in Natural Science gave formal recognition to the position of Science. For some years the scanty class lists bear eloquent witness to the dearth of students. The Reports of the University Commissions show how this dearth was not the only difficulty with which the new subject had to cope. Classics and Mathematics certainly held a privileged position, and it required the steady efforts of men who were looked upon as dangerous reformers to win the firm ground which Science now holds. For instance, at Cambridge H.R.H. Prince Albert, though equipped with the prestige of a Prince Consort and a Chancellor of the University, had to exercise all his tact and influence before the possessors of power there could be convinced that reform was needed.

But Oxford and Cambridge were not to be left in sole possession of the University territory in England and Wales. Durham had been founded in 1832, London University in 1836. Between that date and the end of the century the Royal College of Science was founded in London and fourteen University Colleges were established in the more important towns. Many of these subsequently developed into Universities. At both stages of their career they did incalculable service to the cause of Science in offering stimulating teaching and opportunities of research to many – men and women – who were pressing to enter the realms of new knowledge. But even though there was a bias in favour of Science they were handicapped, as the elder Universities were, by a lack of students. Even those with the enterprise to force their way through the obstacles of their circumstances came often ill-prepared by previous education, and much ability was left untapped. That so much was done under such conditions only intensifies our regret that so much was lost. Genius has a way of saving itself, but it cannot be doubted that a sad amount of the general ability on which educational tone and steady scientific progress depend ran to waste for want of opportunity or on account of misdirection.

4. And now it is the war and its needs that have made us once again conscious of the nation's weakness in Science. But it is for the sake of the long years of peace quite as much as for the days of war that some improvement in the scientific education of the country is required. Just now, everyone is prepared to receive Science with open arms, to treat it as an honoured guest in our educational system, and to give it of our best. Just now, it seems almost unnecessary to take action to ensure against any relapse into the old conditions, but experience of the past shows us that temporary enthusiasm needs to be fortified by some more binding material. Good will is much, but good will weakens, and we must not sacrifice the future to our fears or even to our love of liberty in educational matters. It ought not to be beyond the wit of man to devise a scheme of education that will be durable, yet elastic; a scheme that, while securing that every child should be equipped with a knowledge of Science, will not cramp the teacher by a syllabus or even by a rigid tradition.

Some of the advocates of scientific training have damaged their cause by claiming too much for their subject and by seeming to depreciate the value of the literary studies which had tended to monopolise the attention of the ablest boys who enjoyed secondary education. To many Greek and Latin have seemed enemies who, from having occupied the educational ground betimes, have been able to dig themselves in and to hold an almost impregnable position, due not to their merit as educational instruments but to the accident of priority. There is truth in this, but we do not think that the surest method of victory is to be found in the overstatement of the merits of Science or the depreciation of the value of Classics. Some of the ablest minds have received from their classical instruction enduring gifts that have been of great service to the State and of great refreshment to their possessors. It is our belief that a better service can be done and a like refreshment gained by those whom we hope to see educated on the wider lines laid down in our Report. The humanising influence of the subject has too often been obscured. We are however confident that the teaching of Science must be vivified by a development of its human interest side by side with its material and mechanical aspects, and that while it should be valued as the bringer of prosperity and power to the individual or the nation, it must never be divorced from those literary and historical studies which touch most naturally the heart and the hopes of mankind.

5. There can be no need now to labour the important part that Science should play in our education, but memories are short and it may be well to register in formal words for future comfort if not reproach, what all would readily grant at this moment. It is not possible to give an exhaustive account of the scope of Science but it is not superfluous to point out that it has several distinct kinds of educational value. It can arouse and satisfy the element of wonder in our natures. As an intellectual exercise it disciplines our powers of mind. Its utility and applicability are obvious. It quickens and cultivates directly the faculty of observation. It teaches the learner to reason from facts which come under his own notice. By it, the power of rapid and accurate generalisation is strengthened. Without it, there is a real danger of the mental habit of method and arrangement never being acquired. Those who have had much to do with the teaching of the young know that their worst foe is indolence, often not wilful but due to the fact that curiosity has never been stimulated and the thinking powers never awakened. Memory has generally been cultivated, sometimes imagination, but those whose faculties can best be reached through external and sensible objects have been left dull or made dull by being expected to remember and appreciate without being allowed to see and criticise. In the science lesson, the eye and the judgement are always being called upon for an effort, and because the result is within the vision and appreciation of the learner he is encouraged as he seldom can be when he is dealing with literature. It has often been noticed that boys when they begin to learn Science receive an intellectual refreshment which makes a difference even to their literary work. It is possible to imagine a time when the obstacle to progress in scientific education might be the attitude of scientific teachers, but that time is far distant and it is hard to believe that the teaching of a subject whose life depends on discovery can for long be sterilised, as has

been at one time or another instruction in almost all the other branches of human knowledge.

Too few parents of this generation can satisfy their children's curiosity about the wonders of the heavens, the movement of the planets, the growth of plants, the history of the rocks, the dawn of animal life, the causes of tide and tempest.

How necessary Science is in war, in defence and offence, we have learnt at a great price. How it contributes to the prosperity of industries and trade all are ready to admit. How valuable it may be in opening the mind, in training the judgement, in stirring the imagination and in cultivating a spirit of reverence, few have yet accepted in full faith.

A nation thoroughly trained in scientific method and stirred with an enthusiasm for penetrating and understanding the secrets of nature, would no doubt reap a rich material harvest of comfort and prosperity, but its truest reward would be that it would be fitted by 'an ample and generous education to perform justly, skilfully and magnanimously the offices both private and public of peace and war'.

1.5 THE RESEARCH COUNCILS AND A MINISTRY OF RESEARCH

1.5.1 The Haldane Report and the Research Functions of Government

From Ministry of Reconstruction, Report of the Machinery of Government Committee, under the chairmanship of Viscount Haldane of Cloan, OM, Cd. 9230, 1918.

The Ministry of Reconstruction was set up in the middle of the first world war and lasted only until the middle of 1919. In this short period a number of committees were formed to survey nearly every aspect of the country's life. Although little of positive worth came from all these labours, it was, in A. J. P. Taylor's words, 'a startling recognition of the obligations which the State owed to its citizens and a first attempt to bring public affairs into some kind of rational order'.* The Haldane Committee, although predating

* A. J. P. Taylor, *English History 1914–1945* (Penguin Books, Harmondsworth, 1970), p. 132.

the forming of the Ministry, became one of its investigatory bodies. Ch. IV of Part II of their Report is concerned with research and information and the extracts below come from this. Their statement of the constitutional position of the Research Councils has long been the seminal one, and although no government action may be positively traced back to the Report it is remarkable for its pre-science.

56. In choosing a generic term to describe Intelligence and Research work not supervised by an administrative Department, we have avoided the expression 'Centralised Services', because we regard the relation between the departmental and the general work in these spheres as a problem in correlation rather than in demarcation. For such expressions as centralised services and demarcations of function have acquired a secondary meaning explicitly opposed to the attitude of mind which we intend to suggest. Intelligence and Research work for general use succeeds only insofar as it helps all those whose interests it touches; and correlation cannot imply the exclusion of those engaged in Intelligence or Research work, whether in a departmental or other capacity, from concern with the relevant activities of others who are pursuing the common task.

57. We are clearly of opinion that the expansion of Intelligence and Research work for general use would not conflict with our view that all Departments which have already made distinct provision for Intelligence work should continue to do so, and that many which have not might do so with great advantage; that most Departments must continue to provide themselves with the organisation which they need for the collection and collation of statistical material acquired in the course of their administration; and that many Departments must retain under their own control a distinctive organisation for the prosecution of specific forms of research.

66. As regards the methods to be adopted for conducting inquiry and research in any branch of knowledge, so far as it is determined that the work should be carried out under the supervision of a general organisation, and not under that of an administrative Department, we think that a form of organisation on the lines already laid down for Scientific and Industrial Research will prove most suitable.

67. We may summarise the advantages of such a form of organisation as follows:

(a) It places responsibility to Parliament in the hands of a minister who is in normal times free from any serious pressure of administrative duties, and is immune from any suspicion of being biased by administrative considerations against the application of the results of research.

(b) It gives any authority established under it a jurisdiction which not only extends over the whole United Kingdom, but also facilitates the establishment of relations with research bodies in the Dominions and Colonies, and in India, which find it not unnatural to look upon any organisation under the Privy Council as a body affiliated to themselves.

(c) It leaves open the question of the devolution of administrative services to which we have alluded in Part I of this Report.

68. An expansion of this form of organisation could readily be brought about by the constitution of new Advisory Councils of persons with special knowledge and experience of the subject-matter standing referred to them. Each Advisory Council in its turn would maintain a close connexion with the administrative Departments concerned with that subject-matter, by means of the appointment of Assessors, the use of Special Investigation Committees, and the less formal communications which would, no doubt, be found requisite to supplement these arrangements.

69. It would not be practicable for us to attempt to enumerate the various directions in which the development of work conducted on these lines might proceed. This will depend upon the financial resources available, the advance of knowledge, the opening up of new problems, and the progress made in correlating the research work of administrative Departments with such an organisation as we have described.

70. Clearly the organisation might develop so as to include research in medicine, the diseases of animals, fishes, and plants, in agriculture, botany, and forestry, in problems connected with surveys (including questions of geodesy, geology, meteorology, and oceanography), in the preparation and use of statistics, and

possibly in matters of education and of historical and political science.

71. It will be seen from the Chapters of this Part of our Report dealing with Education and Health that we recommend the reconstitution of the Geological Survey and the Medical Research Committee on these lines. We have not found it practicable in other cases to enter upon such a detailed study of the readjustments which a similar procedure would involve in the position of the bodies by which research is now being conducted as to justify us in submitting specific recommendations in regard to them. But if the principles suggested in Part I and in this Chapter of our Report were accepted, their application to particular provinces of knowledge could be left with confidence to the authorities concerned with each subject-matter from time to time.

72. To envisage any such development is, however, to reveal the difficulties which may supervene in the position of the Lord President of the Council as the responsible minister.

The advantages of placing the authority responsible for Scientific and Industrial Research under the Privy Council have already been alluded to. It was, no doubt, fortunate to have at hand, in the initial stages, a form of organisation through which these advantages could readily be secured. But at the moment when this Report is written, the Lord President is also a member of the War Cabinet and Leader of the House of Lords. It may, it is true, be possible to regard the circumstances which place the burdens of the first of these offices upon the Minister responsible for Scientific and Industrial Research as wholly exceptional. But it must always be not unlikely that the Lord President will be required to take the Leadership of the Upper House.

So long as there is attached to the Committee of Council concerned with Research over which he presides no more than a single Advisory Council, it will, no doubt, be possible, though not easy, for the Lord President to discharge his responsibility for Research with success.

73. If, however, as seems probable, it is found expedient in course of time to establish Advisory Councils for a substantial number of distinct branches of research, it may well prove impracticable for the Lord President to combine the supervision of a range of work so wide and of such high importance with other exacting duties. The time will then have come when the principles

implied in the organisation of work of this kind under the Privy Council have won general acceptance, and the arguments for maintaining that form of organisation, insofar as it lays Ministerial responsibility upon the Lord President, will have diminished in force.

74. It may, therefore, not be premature to anticipate that the distinctive character of the organisation of Intelligence and Research for general use; the proper scope of such an organisation; and its potential relations with analogous organisations throughout the Empire, could thenceforth all be maintained by a Minister specifically appointed on the ground of his suitability to preside over a separate Department of Intelligence and Research, which would no longer act under a Committee of the Privy Council, and would take its place among the most important Departments of Government.

1.5.2 Further Support for a Ministry of Research

From H. A. L. Fisher, A Ministry of Research: Memorandum by the President of the Board of Education, February 1919, Public Record Office, DSIR 17/119.

Among those who supported the conclusions of the Haldane Report's study of research in government was H. A. L. Fisher, President of the Board of Education, who, in this unpublished Cabinet memorandum, put his own reasons for supporting a separate Department of Intelligence and Research.

1. I wish to bring before the notice of the Cabinet some considerations bearing upon the establishment of a Ministry of Research, an issue, which quite apart from the recommendations of the Committee on the Machinery of Government, must necessarily force itself before long upon the attention of Parliament and the public by reason of the large sums which are now being spent, and of the still larger sums which will shortly be required for the advancement of scientific discovery in every department of knowledge.

2. It will be generally recognised that while each of the great administrative departments of Government should possess an intelligence and research section, the formation of a central organ for scientific discovery and information is also essential.

3. It is also clear that, whatever other characteristics this Central Department of Research should possess, it should be framed with reference to three principles: It should be unembarrassed by the burden of ordinary administrative work; it should be empowered to promote research for the whole United Kingdom; and it should be able easily and naturally to coordinate its scientific activities with the analogous activities in the Dominions, the Colonies, and the Empire. In a word, it should be an organ of intelligence for the whole Empire, unburdened by the routine of current administration.

4. For these reasons there is much to be said for accepting the Committee of the Privy Council for Industrial and Scientific Research as our central organ for research and intelligence. It conforms to each of the three principles which have been mentioned.

5. On the other hand, the arrangement under which the Lord President of the Council is the responsible Minister for Research is clearly wrong in principle. The Lord President is generally one of the older statesmen. He may be, and often is, unversed in science and out of sympathy with scientific development. He may again be well qualified to discharge the duties of the post, but heavily burdened by other work, as is notoriously the case with the present holder of the office. If the Department is to exercise its legitimate influence and to realise all the advantages of which it is capable, then it should be presided over by a Minister who is chosen because of his peculiar fitness to deal with this special type of work. Any Lord President of the Council may be such a person, but he clearly does not acquire the necessary qualities *ex officio*.

6. It may be urged that the Lord President may delegate all or part of this work to some other Privy Councillor, e.g. the President of the Board of Education, but this too is an unsatisfactory arrangement, for no Minister who is ultimately responsible for a Department can delegate his responsibility in full, and shared responsibility must make in the end for bad and languid administration. I have been attempting to cope with the work of the Committee for the past few months, but I am fully conscious that, owing to the heavy calls of my own Department, I have not been able to give to industrial and scientific research the attention which it deserves.

7. For these reasons it would, in my opinion, be well that an

early opportunity should be taken of giving a statutory basis to the position of the Privy Councillor who is entrusted with the actual work of directing the Department of Research. His title, emoluments, and duties should be defined by Parliament. Just as there was a Vice-President of the Council for Education, so there should now be a Vice-President of the Council or Vice-Chairman of the Council for Research.

8. I am not prepared to say that it would not be possible, as it would undoubtedly be economical, to combine this post for the present with one of the lighter offices of the Government, but I am convinced that in the course of a few years the work of the Department will require the full time of a Minister. Here, for instance, is an outline of the possible branches of a Department of Research, each branch to be organised under an Advisory Council of its own, and many of the branches to be connected with Research Boards in executive control of research industries:

I. – *Scientific and Industrial Research.* – This branch already deals with the encouragement of pure science in all scientific subjects not already provided for by other State organisations. Further, it deals with industrial research of all kinds not covered by other organisations, such as the Development Commission or the Medical Research Committee. The proposed future organisation for aeronautical research would come under the purview of this branch.

II. – *Geophysics Research.* – This branch will deal with the pure sciences related to geophysics, and on its applied side to the geological surveys, including the Geological Museum, the Meteorological Office, and in course of time no doubt also with geography, oceanography, and geodesy.

III. – *Medical Research.* – This branch would deal with the present and future work of the present Medical Research Committee.

IV. – *Agricultural Research.* – This branch would deal with the pure sciences which have a bearing upon agriculture and with such fundamental applied investigations as would be carried out by, e.g. an Institute of Animal Nutrition and an Institute of Comparative Pathology. Both these institutes would also be closely connected with the branch for medical research, since the nutrition and diseases of animals are now seen to be of great importance in connexion with the nutrition and diseases of man. Neither can be satisfactorily investigated apart from the other.

V. – *Statistical Research.* – This branch would deal with problems of statistics in relation probably to the public well-being, and would for the first time work out a system of vital statistics in close touch with the Medical Research Branch. It would also, no doubt, be in a position to arrange with the statistical branches of the various Government Departments for the presentation in a convenient form of collected statistics of the work of these Departments on a uniform basis. At present no attempt whatever has been made in this direction.

VI. – *Economic Research.* – This branch would deal with the encouragement of research in pure economics, but also with a number of pressing problems in the application of science, such as costings.

9. It will be seen that all the functions enumerated above are concerned with original investigation, and are directed to the discovery of new truth. It has been suggested that the Ministry should also be entrusted with the duty of collecting sociological and political information. It is probable, however, that this function would be best discharged in the first instance by the departmental organs of inquiry – that the Home Office, for instance, should contribute information with respect to the cost of factory legislation abroad; the Board of Education with respect to educational changes; the Ministry of Health with respect to questions of public hygiene. In any case, the creation of effective departmental organs for this type of inquiry is prior in urgency to the creation of a central bureau of sociological information.

10. It is not necessary that the Minister of Research should be a great man of science, distinguished for original contributions to knowledge, but obviously it is desirable that he should not be entirely ignorant of science or out of touch with it. A man of wide scientific interests, acquainted with academic machinery and the scientific personnel of the country, and at the same time gifted with administrative ability, would be the sort of man who might be expected to fill the office to the greatest advantage. The work of the Department will fall into so many branches, each of them highly specialised, that no specialist Minister could hope to form a personal opinion of any value upon the great mass of the recommendations coming up to him. He would have to trust to Specialist Committees, and he ought at least to possess a lively interest in all

the great scientific developments of the time. It may be asked whether the Minister for Research should combine with his strictly scientific duties some of the functions recently delegated to the Minister of Reconstruction. Should the new Ministry, in other words, be asked to skirmish ahead of the Government of the day, to explore the social problems of the future, and to prepare plans for legislation? These functions imply qualities differing from those which properly belong to a Minister of Scientific Research, and I doubt whether it would be wise to burden the Department with the task of framing suggestions for legislation, except insofar as those suggestions may be directly connected with the march of physical science. Experience indeed has shown that the influence of the existing Department of Scientific and Industrial Research on the industries, and even more perhaps of the administrative departments of State, depends upon its detachment from all problems of exploitation and from administrative policy. The Department is coming to be recognised as completely unbiased, and concerned only with the discovery of scientific truth and its dissemination. The application of the results obtained rests either with the industries concerned or with the Department for whom the investigation has been made. I think this is an important principle, from which it would be dangerous to depart.

1.5.3 The Formation of the Medical Research Council

From Dr Christopher Addison, Memorandum on the Future Organisation of Medical Research, March 1918. Reprinted as an appendix to the Annual Report of the Medical Research Council for 1950–51, Cmd. 8584, 1951–2.

Following the end of the war the Medical Research Committee was reformed as the Medical Research Council, following the recommendations of Dr Christopher Addison (then Minister of Reconstruction) and the Haldane Committee. In this memorandum, Addison, who had been connected with the Medical Research Committee since his support for Lloyd George during the passage of the National Insurance Act, set out the reasons why he felt medical research should be transferred to a Committee of the Privy Council.

Practical Reasons for Reorganisation

5. The reorganisation of medical research work is a problem which extends beyond the Medical Research Committee itself. The Local Government Board, the Board of Control, the Ministry of Munitions, the Colonial Office, and the Departments concerned with national defence, are all at the present moment spending money on medical research, some of them using the Medical Research Committee as an agent for placing particular pieces of research in the best hands, and some of them acting independently. It has never been proposed that the Medical Research Committee, or any similar body which succeeds it, should prevent all other Departments from doing anything in the nature of medical research, and the question therefore is whether the valuable influence of the Medical Research Committee in preventing overlapping inquiries and using the best scientific men can be maintained most effectively if it is organised on lines similar to those at present laid down, or under some different arrangement.

6. In order that a considered judgement on this question may be formed it is necessary to give some account of the relations which have existed, since the Medical Research Committee was first created, between the Committee and the Minister responsible for Health Insurance. The position is that the Medical Research Fund Regulations lay upon the Medical Research Committee the duty of framing schemes for research. Those schemes are submitted for approval, not to the Joint Committee as a whole, but to the Chairman of the Joint Committee, as the responsible minister, in person. As soon as the minister's approval has been given to a scheme, the Committee are left free to carry it into operation, and the Secretary to the Committee is responsible for seeing that the approved expenditure is not exceeded, and that expenditure is made, within the estimate, upon the proper heads of the scheme.

Although, therefore, the operations of the Medical Research Committee are under the control of the Minister responsible for Health Insurance, so that he would defend the proceedings of the Committee if they were criticised in Parliament, it will be seen that in practice the Minister relies upon the Medical Research Committee to select the objects under which they will spend their income, and to frame schemes for the efficient and economical performance of their work. The Minister has, of course, always received a full explanation of their schemes from the Committee before giving his approval, but he has never sought to control their work, or to suggest to them that they should take one line rather than another, as all Ministers rightly do in the administrative work of their Departments.

7. There is, therefore, an important distinction to be drawn between this research work and all other work within the sphere of the Department, whatever its name; and the judgement of the eminent scientists who are members of the Medical Research Committee as to the value of this undertaking is perfectly clear. In their First Annual Report (1914–15, Cd. 8101, p. 48) the Committee say that they 'venture to acknowledge their indebtedness to the three successive Chairmen of the National Health Insurance Joint Committee under whom they have worked, for having allowed them the most complete freedom, within their constitution, to bring flexible and rapid assistance to the national need on occasions of emergency, with the least possible delay in the motion of constitutional machinery'.

8. It may be asked, however, why, if the relations of the Medical Research Committee with the present Department are so satisfactory, they cannot be left as they are under the Bill establishing the Ministry of Health. The answers to this question are of two quite different kinds. There are, in the first place, some serious difficulties arising out of the present constitution of the Medical Research Committee, which will, in any case, have to be met as soon as the central administration of Health Insurance is altered, as it must be, by the establishment of a Ministry of Health. It is proposed to establish a Ministry for England and Wales only, with some consequential adjustments as regards Health Insurance in Scotland and Ireland. But the Medical Research Fund has from the first been deliberately made a single fund for the United Kingdom. It was necessary to take this course in order to make the best use of the comparatively small amount of money available, and the experience of the Committee has shown that for effective work the committee must be in close touch with the best scientific activities in all parts of the whole Kingdom; in any given piece of research it may be necessary, or highly advantageous, to bring into association with work being done at one university or other centre the investigations by some other worker far remote, and belonging, perhaps, not only to another nation, but even to a different kind of scientific object.

9. In the second place, the independence of the Committee has rested not only upon the particular constitution which was framed for them when they were appointed, but also upon the fact that the Insurance Departments in the six years of their existence have been so much absorbed in putting the Acts into operation and in improving their administration that they have had very little time to devote to health problems in the more scientific sense. It might appear at first sight that when a Ministry of Health is set up, and a more advanced health policy is adopted, as we hope it will be as soon as the Ministry is established, the arguments for keeping the Medical Research Committee closely linked with the new Ministry would be strengthened; but I think that on closer examination the arguments to the contrary are stronger.

10. A progressive Ministry of Health must necessarily become deeply committed from time to time to particular systems of health administration. The Minister of Health at any moment may be appointed by the Government on the ground that he is something of a scientist or takes a special interest in health matters. One does not wish to attach too much importance to the possibility that a particular Minister may hold strong personal views on particular questions of medical science or of its applications in practice; but, even apart from special difficulties of this kind, which cannot be left out of account, a keen and energetic Minister will quite properly do his best to maintain the administrative policy which he finds existing in his Department, or imposes upon his Department during his term of office. He would, therefore, be constantly tempted to endeavour in various ways to secure that the conclusions reached by organised work under any scientific body, such as the Medical Research Committee, which was substantially under his control, should not suggest that his administrative policy might require alteration. The more active the administration of his Department the greater this danger becomes. It is essential that such a situation should not be allowed to arise, for it is the first object of scientific research of all kinds to make new discoveries, and these discoveries are bound to correct the conclusions based upon the knowledge which was previously available, and, therefore, in the long run to make it right to alter administrative policy.

11. Accordingly, any body of men engaged upon scientific research in medicine or any other field should be given the widest possible freedom to make their new discoveries, and to make them

available for the use of the administrative departments. This can only be secured by making the connexion between the administrative Departments concerned, for example, with medicine and public health, and the research bodies whose work touches on the same subjects, as elastic as possible, and by refraining from putting the scientific bodies in any way under the direct control of ministers responsible for the administration of health matters.

12. Further, it must be remembered that, even apart from direct interference by the Minister of Health, the Medical Research Committee, if it were specially attached to his Department, would tend to be too much absorbed in making researches into those problems which appeared at the moment to be of the most pressing practical importance. These problems must, of course, be effectively dealt with in the interests of the good administration of the Ministry of Health. It is for this reason that the Ministry must always conduct some researches through its own staff. The Department must also be in the closest touch with any body, such as the Medical Research Committee, which can give assistance in solving such practical problems. But, while it is essential that the administrative Departments should let the scientific body know what are the practical problems of the day calling urgently for inquiry, the scientific body should not be limited to dealing with the practical aspects of those problems. It has already been found in many cases that an inquiry started with a purely practical purpose has led scientific men into new inquiries, resulting in fresh discoveries which have been valuable for purposes quite distinct from the solution of the original problem. It has been found equally that the solution of a particular problem has often come quite unexpectedly from scientific work in some other direction, that would have been thought at first sight to be wholly remote from it.

13. This freedom to pursue scientific inquiries in any direction which may increase scientific knowledge of any kind is implied in the words of the Act of 1911 which refer to medical research. At the outset of the Medical Research Committee's work it was clearly understood that they should not be limited to inquiring into problems arising out of the current administration of the National Insurance Acts, but that they could inquire into any subject which was covered by the words 'Medical Research'. In fact, as is well known, the main energies of the Committee have been devoted, ever since the outbreak of war, to the investigation

95

of practical problems arising in the course of the work of the Admiralty, the War Office, and the Air Force. It is not suggested that those problems, or the other problems which the Committee had begun to investigate before the war, have no bearing upon the health of insured persons in particular, but the subjects for investigation have always been selected by the Committee on account of their general medical and scientific importance, and not because the Insurance Department thought that these subjects should have the first claim upon the time and funds of the Committee.

14. A further important argument against associating the Committee with a single strong administrative Department is that this course would undermine the confidence at present felt in the Committee by the large number of Departments which have from time to time made demands upon the Committee's services. If the Committee is to continue to be of the fullest service to all Departments which want advanced scientific research in medical subjects to be undertaken on their behalf, it must continue to work in friendly relations with the medical and other officers of all such Departments, and to have free access to the Departmental papers, which may often be of a confidential character.

15. But it is certain that if the Committee were known to be working in specially close relation with a progressive Ministry of Health, and also to be substantially under the control of the Minister, all other Departments would begin to object to using the Committee and giving it full information, and would do their best to conduct the whole of the medical research which they required through their own officers. This would prevent any single body, such as the Medical Research Committee, from having under their view the whole of the medical research which was being done on behalf of the Government, and would make impossible the proper distribution of the work between the separate Departments and the medical research body which should be the common helper of all Departments.

2 Central Organisation: The Interwar Years

The years from 1920 to 1939 were years of triumph and tragedy for science in government, the universities and industry. There are many ways of describing the changing characteristics of science and the scientific community during this period; this chapter isolates four such.

First, there is the contribution to scientific policy made by the 'new generation' of scientific administrators associated with the research councils and with the scientific work of the departments. For the first time in the history of public administration there was a group of professional scientific administrators whose policies, attitudes and prejudices influenced every aspect of science policy: men such as Sir Frank Heath of the DSIR, Sir Walter Fletcher of the MRC, and above all, Lord Balfour, who, as Lord President of the Council, made an outstanding contribution to an emergent 'scientific policy' which has tended to be neglected in the context of his other political activities [2.1.1, 2.1.2].

Second, there is the history of the scientific institutions themselves. The economic retrenchment of the early 1920s, followed after a brief respite by the international crises of 1929–30, threatened to curtail and frustrate the first steps of the DSIR and MRC. The DSIR, whose scientific priorities and policies had, ironically, been determined by the national problems of wartime, such as fuel, food and timber shortages, fought hard against this retrenchment in the early 1920s. But it was not until 1926 that there was any appreciable improvement in the DSIR's financial position. Important building programmes were postponed year after year, and research was confined to that which was strictly in the national interest, such as that on fuels, building materials and grants to students for their research. In short, priorities were determined closely by national economic and social policies. The MRC, on the other hand, did not suffer in the same way. The advantage of a five-year grant, although bringing with it certain disadvantages towards the end of the period, permitted a greater degree of flexibility and coordination in

97

planning research programmes. The 1920s, therefore, were not the years of progress which the end of the war had promised. Extracts from the annual reports of the research councils, including the Agricultural Research Council, established in 1931, illustrate some of the major policy issues during the period [2.2.1–2.2.7].

Third, one of the more neglected characteristics of the period, but one which constantly runs through the planning of science in these years, was the concept of the 'coordination' of research. This was both a result of the constant Treasury demands for economy and an in-built feature of the consolidation and rationalisation of research programmes. Coordination of the defence research services and the civil research programme was the subject of several cabinet committee investigations [2.3.1]. The aim of coordination was also associated with the wider aim, envisaged by Lord Balfour, of using science to solve social and economic problems, especially those associated with imperial development, such as the eradication of diseases and pests. The Committee of Civil Research, essentially Lord Balfour's creation, and a much underrated exercise in scientific policy, united the twin themes of coordination and the expert use of scientific knowledge, both in the Empire and at home [2.3.3 – 2.3.5]. Some of the work of the Committee of Civil Research was carried on throughout the 1930s by the Standing Committee on Scientific Research of the Economic Advisory Council. With the growing threat of German militarism there was a renewed effort to provide a coordinated effort uniting research in the fighting services. The anxiety of scientists such as Ernest Rutherford and W. H. Bragg to avoid the mistakes of the first world war and to prepare as fully as possible for the organised employment of scientists in the event of a new war, was very apparent in their attempts from 1938 onwards to set up research coordinating committees [2.3.6, 2.3.7].

The final theme in this chapter concerns the scientific community. Against a background of international violence and political extremism one of the most important scientific movements was born in the early 1930s – the debate on the freedom and planning of science. The debate can be said to have been sparked off at the International Congress on the History of Science in London in 1931 when the Soviet delegation argued that the results of research were essentially a byproduct of national social and economic conditions. The idea of scientific research as a tool of human advancement found enthusiastic advocates in Britain, among them leading scientists and writers such as J. D. Bernal, Lancelot Hogben, Julian Huxley and J. G. Crowther. Their demands that science should be seen 'as a social activity and not as something apart from the rest of human life and interest' were articulated in *The Frustration of Science*, a collection of essays published in 1935 in which P. M. S. Blackett argued that the traditional distinction between pure and applied science was invidious and misleading [2.4.2]. *Nature*, under the spirited editorship of Sir Richard Gregory, supported their demands for the 'scientific spirit and method in the shape of careful planning' [2.4.1]. In August 1938, reflecting the growing concern with the principle of social responsibility in science, the British Association set up a special Division for the Social and International Relations of Science. In 1939 J. D. Bernal published *The Social Function of Science* which, more than any other single document, offered a new blueprint for the organisation of scientific research [2.4.4]. There were other scientists, particularly Michael Polanyi and J. R. Baker [2.4.5] who saw in this movement a new form of totalitarianism: an attempt to tie research down to the utilitarian demands of the state. Through the Society for Freedom in Science they argued for the independence of the scientist and his right to obey the dictates of his imagination and conscience. It is a debate of particular relevance to the international community of scientists today.

2.1 SCIENTIFIC ADMINISTRATORS

2.1.1 Tizard at the DSIR

From Sir Henry Tizard's Haldane Memorial Lecture, 'A Scientist in and out of the Civil Service', delivered at Birkbeck College, London, 9 March 1955, pp. 7–8, 10–11.

In his biographical study of Sir Henry Tizard (Methuen, London, 1965, p. xv), Ronald Clark wrote that 'Sir Henry Tizard's life is a record of one man's influence on defence science between 1914 and 1951, a period which saw it transformed to the decisive factor in a nation's survival'. Tizard represented many different aspects of the statesmen of science. He was closely connected with the development of test-flying and radar; he was 'one of the few who knew just which Civil Service button might be pressed, at what moment, by whom, with the greatest effect'. The following extract briefly illustrates his early work as a senior civil servant at the DSIR. His first responsibility was to organise the work of the physics, chemistry, engineering and radio Coordinating Boards [2.3.2]. In 1927 Sir Frank Heath retired and Tizard succeeded him as secretary. His work at the DSIR was characterised by his energetic conviction that 'a nation's industrial success ... can only be achieved by the planned utilisation of scientific knowledge' (Clark, *op. cit.*, p. 76). [See also 6.6.]

I joined the Department of Scientific and Industrial Research in 1920, and so became a member of the Civil Service for the first time. But it was not my first experience of the Civil Service. At the end of 1917, when I was in the Royal Air Force, I was transferred to Headquarters in London to act as assistant to Bertram Hopkinson who was then in charge of all experimental work for the Royal Air Force. He was a really great man whose death in a flying accident in the summer of 1918 deprived the nation of one whose leadership would have been invaluable in the years between the wars. I learnt much from him; but I did not emerge from that employment with any great respect for the manner in which the Civil Service dealt with highly technical matters. The best, or perhaps the worst, example of what I mean is shown by the following story. A file about a new invention arrived in my office one day. I have tried in vain to disinter it from its grave – it was probably cremated long ago – so it may be pure imagination on my part to say that it started with a minute written in green ink over the well-known initials W.S.C. It certainly was not marked 'Action to-day' because it had gone through the hands of many able administrators before it reached me. The invention was that of a mysterious liquid one drop of which added to a gallon of water would make a first-class aviation fuel; and the sum asked for the secret was very large. The able administrators had enlarged on the diminishing stocks of petrol, on the rate of consumption, the sinking of tankers, and so on – the general conclusion, implied if not expressed, being that if only half what the inventor claimed was true it would be well worth-while paying the sum asked for the secret. I returned the file to its source with what I think may have been the shortest minute ever written to a Minister. I did not expect to hear of the invention again, but much to my amusement it turned up soon after I had joined the D.S.I.R. This time the sponsor was Mr Horatio Bottomley, MP, who was encouraged by ministers to approach the Department. He must have been very disappointed when the Department refused to make an official test – and the price of oil shares remained steady. This is by no means the only instance I could quote of 'inventions' being taken seriously by intelligent men of affairs which I should have thought a schoolboy would reject off-hand. Nothing has illustrated to me more vividly the immense gap that still exists between educated men of scientific and literary upbringings. It is a gap that must be filled somehow.

I enjoyed my time at the Department of Scientific and Industrial Research. It was a new venture of Government, and I think that I have always enjoyed myself most, and perhaps have been more useful, in the nursery stage of new developments, before they have grown up, so to speak, and are out in the world and respected by all. The Advisory Council of my days was an exceptionally brilliant group of men, presided over by the genial and wise Sir William McCormick. The three members who made the most impression on me, in their varying ways, were Sir Joseph Thomson, Sir Richard Threlfall and Sir William Hardy. The last named combined to a greater degree than anyone else I have known great distinction in pure science and good judgment and flair in its application. I remember him putting his head round the door of my room one day and saying: 'You know, this applied science is just as interesting as pure science, and what's more it's a damned sight more difficult!' He must have had some human trouble that day.

My nine years at the Department soon passed. My growing familiarity with the leaders of the Civil Service bred a respect which has been enhanced by more recent experience. But I was, and still am, much more critical about the detailed control of the Treasury which is exercised through the Establishment Branches. I did not think that this system was suitable for the financial control of Research Departments. I thought that it called for the expenditure of too much time and money on unnecessary administration. Towards the end of my time I was discussing informally with the Treasury the possibility of treating the Advisory Council in the same way as the University Grants Committee. Under this scheme the various Research Establishments under D.S.I.R. would have enjoyed a large measure of autonomy, their grant from public funds would be settled for five years, at the end of which period they would be judged again on their merits, and the Advisory Council and a small headquarters staff would be set free to nurse new developments through the initial stages, and to study what I have since called the strategy of applied science. When I left in 1929 these discussions went no further; but I am glad to see that a step in the right direction has recently been taken to provide for the continuity of the Department's policy over a period of years.

The position of twenty-five years ago was that the principle that it was one of the duties of Government to promote scientific research in the interests of the nation was accepted by all political parties. This was a revolution in political thought, mainly due to the influence of Haldane, reinforced by the experiences of war. The chief change since then has been one of scale and range. The total net expenditure of the D.S.I.R. when I left in 1929 was £500,000. It is now over £6 million. Even after allowing for the change in the value of money, this is a large increase in twenty-five years. Other Government research organisations show a similar increase. University departments of science are well endowed – and in addition a very large expenditure is now incurred on the civilian uses of atomic energy. Haldane would have no reason to complain of the lack of research nowadays – nor of the apathy of Parliament. Personally I think that this enthusiasm carries with it certain dangers, and I should like to see it tempered by more well-informed criticism. The fact is that there are not enough scientists of the right calibre to tackle all the many problems for which a good case can be made out. Indeed, there never will be, in a properly balanced society. In Government research the broad problem is not simply to secure the services of a fair but not excessive proportion of the best men available, but to use them in the best way at any given time. We must avoid the mistake of rushing into a programme of research to meet a need without studying beforehand the real nature of the need that has to be met. That is why I say that the study of the strategy of science is now of great national importance. It is as easy to waste money on research as on anything else; especially if it is someone else's money.

2.1.2 Lord Balfour, Statesman of Science

From Lord Balfour, Memorandum on the Finance of Research, 1926, Balfour Papers (British Museum, Add. Mss. 49704), ff. 53.

Arthur James Balfour, first Earl of Balfour (1848–1930), occupies a unique place in the history of British science policy. He came from a family of marked scientific interests (his uncle, Lord Salisbury, was famous for his private laboratory at Hatfield House). His political career furnishes many examples of political expertise promoting scientific interests. In 1902 he was instrumental in creating the Committee of Imperial Defence (CID) a consultative council of statesmen to advise and coordinate imperial defence strategy. In 1915 he set up the Board of Invention and Research – an expert scientific committee designed to improve submarine detection methods. In 1925 he borrowed the structure of the CID for the Committee of Civil Research which exercised analogous functions for research. He was Lord President of the Council twice, 1919–22 and 1925–9; his responsibility brought him into direct contact with the DSIR and the MRC. It was here that his genius for reconciling science and politics flourished. He was chairman of the MRC from 1925 until his death. Perhaps the greatest tribute paid to him came from Sir Frank Heath, quoted in Lord Rayleigh's biography of him, *Lord Balfour in His Relation to Science* (Cambridge University Press, 1930, p. 46):

The Lord Presidency used to be considered as a general utility office. He converted it into a Ministry of Research. The idea was not born in his fertile brain, for a Committee of the Privy Council for Scientific and Industrial Research and a similar Committee for Medical Research had been established during the war and Lord Haldane's Committee on the Machinery of Government had recommended the creation of such a Ministry. But Lord Balfour it was who turned an experiment which many thought destined to disappear with other wartime devices into a reality which is now generally recognised as a permanent and essential part of modern government. His unparalleled prestige in the political and intellectual worlds, his liberation from the rough and tumble of party politics, were favourable circumstances, but his abiding faith in the power of science to promote the happiness and wellbeing of man, his enthusiastic interest in the advance of knowledge, his sympathy with the scientific outlook and with young people, and his long experience of the way in which things have to be done in Great Britain, were the decisive factors.

The following memorandum, submitted by Lord Balfour to the Treasury in defence of the DSIR's expanded estimates for 1926–7, illustrates Balfour's belief in the wisdom of state investment in research.

There is one inevitable difficulty in forming a sound judgement on national expenditure on research, namely that in every research scheme there must always be an element of conjecture. Research by the very meaning of the term is an attempt to penetrate the unknown and until the attempt is made there can be never any assurance that the resulting knowledge will be adequate to the effort made to attain it. Unfortunately it is just at the moment when risks of this kind need most certainly to be run that the temptation to avoid all risks is most powerful. This is the vicious circle which private industrial enterprise always gets entangled in in times of depression and discouragement. I believe it is in part because we are entangled in it at this very moment, while the Americans are not, that our countries stand at such different levels of confident prosperity. The great American undertakings scrap recklessly – investigate boldly and spend lavishly on every hopeful scheme for increasing efficiency of output. The Britisher, suffering under the discouragement with which we are all too familiar, is forced to look around for economies and finds them all too easily in cutting down the researches required by his business and improvements in replacement of his old obsolete plant. These probably are the easiest of all economies but they may well be the most costly in the long run . . .

Now if this be true, or anything like the truth, the activities of the DSIR are of greater importance in times of industrial depression than in any other, and alterations of policy involving serious diminution of its efficiency ought to be most critically examined. The doctrine that the pursuit of pure science should be left to the Universities, Learned Societies or private investigators and its practical applications to the industries interested is arguable, though I think erroneous. What surely is not arguable is the policy of establishing an organisation for scientific and industrial research, enlisting in its service the best brains in the country, starting schemes through it, which from the nature of the case require precision in design and continuity in execution, and then after all this effort bringing them to an ineffectual conclusion – *not* because they are bad but because the depressed condition of industry conclusively shows that they are specially necessary.

2.2 THE DEVELOPMENT OF THE RESEARCH COUNCILS

2.2.1 Retrospect of the First Five Years of the DSIR

From the Report of the Committee of the Privy Council for Scientific and Industrial Research for 1919–20, Cmd. 905, 1920, pp. 11–13, 18–19.

In their first postwar Annual Report the Advisory Council looked back over their first five years' work.

The present report covers the fifth year of our work since our appointment under the Order in Council of 28 July 1915, and it appears proper that we should take this occasion of reviewing what has been achieved during the formative period of our existence. When we held our first meeting on 17 August 1915 the war was in its second year, and it was apparent that if victory was to be secured it would call for the organisation of all the resources and of the whole adult population of the realm. The Allied offensive on the west had failed. The Russians were in retreat. The Munitions of War Act had received the Royal Assent on 2 July and the first step was taken towards universal military service by the passage of the National Registration Act which became law on 15 July. From the date of our beginning until November 1918, that is to say, for more than three-fifths of the period under review, the intensity of the national effort towards the single military end was month by month accentuated and increased. Yet the Advisory Council and the Department which, established a year later, slowly grew up round it, were not created as part of the machinery for war. They made their contribution to the common need like every other national organisation, but they were brought into existence to remedy deep-seated short-comings revealed by the war, and the full effect of their activities could not be looked for until after peace had come.

The Advisory Council were at that time directed by Your Lordships 'to frame a programme for their own guidance in recommending proposals for research and for the guidance of the Committee of Council in allocating such State funds as may be available. This scheme will naturally be designed to operate over some years in advance, and in framing it the Council must necessarily have due regard to the relative urgency of the problems requiring solution, the supply of trained researchers available for particular pieces of research, and the material facilities in the form of laboratories and equipment which are available or can be provided for specific researches. Such a scheme will naturally be elastic and will require modification from year to year; but it is obviously undesirable that the Council should live "from hand to mouth" or work on the principle of "first come first served", and the recommendations (which for the purpose of estimating they will have to make annually to the Committee of Council) should represent progressive instalments of a considered programme and policy.' It is now more or less generally recognised that an industrial people cannot hope to survive, either in peace or in war, without intensive application of science to all their activities: to their way of life in ease and at work, to their education, to the raw materials and processes of their manufactures and industries, to their local and central administration, to the winning and conservation of the natural resources of their country, and not least to their means of communication. It is not so generally appreciated even now that unless the creation of new knowledge is similarly encouraged the application of science to our growing needs must soon reach its limit. In 1915 neither of these truths was commonly apprehended and it was our first duty to think out a plan which would make the utmost use of the means already in existence, bring them as far as possible into cooperation, supplement or strengthen them, and add new machinery only where we found no other suitable agency available for our purpose.

Our programme has gradually been defined during the past five years and can now be laid out with considerable precision. It falls under four main heads: (i) The encouragement of the individual research worker, particularly in pure science; (ii) the organisation of national industries into cooperative research associations; (iii) the direction and coordination of research for national purposes; and (iv) the aiding of suitable researches undertaken by scientific and professional societies and organisations. In the following paragraphs we give a short review of what has already been done in each of these spheres.

Apart from the educational side of their programme, the Advisory Council laid special emphasis on five areas of national research: fuel, food, building materials, the work of the National Physical Laboratory and the work of the Geological Survey. In a later paragraph the Council stressed the urgency of the fuel question which had precipitated the creation of the Fuel Research Board:

When we began making our plans for the organisation of the industries of the country for scientific research, it became evident to us that if the scheme for cooperative research in the several industries were to be a permanent success, provision must also be made for dealing with certain fundamental problems which were of such wide application that no single industry, however intelligent or highly organised, could hope to grapple with them effectively.

FUEL A BASIC PROBLEM

The first of these basic problems was fuel. In this country coal is and must remain the principal source of the heat, light and power without which no modern industry is possible. But the right or wrong use of coal affects not only every industry, it has an intimate bearing, through the Navy, the Air Force, and even through the Army, upon the national defence, and it directly influences the comfort, health and general well-being of every man, woman and child in the land. The use of fuel to the best possible advantage, the application of each available kind of fuel to its appropriate purposes, and the discovery and production of new forms of fuel, are obviously national problems and as such can most equitably be attacked at the cost of the taxes to which each of us contributes in his degree. Special aspects of these large questions may properly be entrusted to particular localities or to individual industries or firms, but the wider problems seemed to us to be clearly matters for the State. We therefore recommended Your Lordships to establish a special organisation for the purpose, and the Fuel Research Board was appointed under the directorship of Sir George Beilby, FRS.

2.2.2 The Medical Research Council and the Nation's Food Supply

From the Report of the Medical Research Council for 1926–7, Cmd. 3013, 1928, pp. 22–3.

In the following extract an account of the MRC's early activity in one of the most important and successful areas of its work is given: nutrition.

Nutritional Studies in General

There are other vitamins essential for the health of the community besides the two which have just been discussed. Studies of these are being actively pursued with help from the Council, as the present Report will show. The whole group of studies is under the supervision of a special Committee. This is very costly work, because it involves innumerable animal feeding experiments conducted with scrupulous and unremitting attention. The Council give as liberal support to this work as their limited resources allow. In 1921 they expressed their opinion that this kind of work is a good example of what it is highly proper to support from public funds. They then explained that this development of physiology had originated in this country, but its progress had been crippled at the beginning – which may be placed at about 1912 – and for several years afterwards, by the inability of Universities and other private institutions to provide the necessary funds. The Government grant-in-aid for medical research had allowed more rapid and fruitful increase in knowledge to be gained by workers in this country after a period in which it had seemed that the whole subject must pass after its origination here to development in other countries. They explained that if the war services alone of this kind of work were reckoned, it would have repaid very many times over all that the State had spent upon it. Its value to the nation in time of peace has become in every subsequent year more apparent, affecting as it does at a thousand points the food of the population, its choice, its preparation in manufacturing processes and in the kitchen, and its production by agriculture. To agriculture it has its own importance, too, for the proper nutrition of livestock.

Valuable as they are, however, these studies of vitamins are but one part of the nutritional science of which indeed we are now only seeing the dawn. Modern studies of the nutritional value of inorganic elements or their salts are indicating that here again, as with the vitamins, we are concerned with another science of the 'infinitely little'. Vitamin actions are found to be related in vital ways to subtle factors of balance between them, and again between them and other factors in diet. New knowledge is coming of constituents of diet which appear able to nullify the action of vitamins. Qualitative problems of all these kinds are linked, of course, with the quantitative problems of nutrition. Last year the Council appointed a small representative Nutrition Committee to bring under review all these different lines of study with a view to framing a comprehensive and coordinated programme of work. The Committee are attempting to promote and extend studies of the mixed qualitative problems of the kind just indicated. They are attempting to frame, in cooperation with the agricultural interests, a chemo-geographical survey of the inorganic constituents of diet available in different parts of the country and their relation to the distribution of disorders in health, human or animal. They have also under consideration other sets of problems upon the quantitative side which call for solution. One is that of the so-called 'man value'; what are the factors to be rightly used in referring quantitative dietetic estimates to a single man-power standard by due allowance for age and sex? There is reason to believe that the factors at present used in calculations of dietetic investigators of this and other countries undervalue the requirements of children. Another problem is that of the actual energy requirements of the adult, and here the standard accepted for the average adult man is probably too high. Connected with this is the conundrum of 'luxus' consumption. What is the biological value, if any, of food taken in excess of the actual needs of the body as judged by the energy expenditure? These are all questions of the greatest importance to a nation whether at war – as all European experience vividly showed – or whether at peace. For no country are they more important than for Great Britain, with a crowded industrial population upon an island. The Council cannot hope to contribute effectively to their solution without extended resources of their own and the cooperative effort of other interests.

If we turn from these broad national issues to the immediate medical problems of the day, we see again the vital importance to the public, and to the medical profession who serve them, of accurate dietetic knowledge. The few examples already given illustrate this enough. But indeed the knowledge we have so recently gained only serves to illuminate the darkness of our present ignorance. In hardly more than a dozen years disorders believed to be specific diseases, such as beri-beri, scurvy, and rickets, have become assignable to dietetic errors. We cannot guess what new relations to diet or disease, or of liability to disease, are waiting to be revealed. In many of the simplest matters no physicians, or surgeons, can give advice based on scientific knowledge, even if they hazard opinions based on earnest but uncontrolled observations. It is not yet scientifically known, for instance, how much water a man should drink in the day, or what is the best kind of bread to eat. It is for those who have the most reason to know the public value of this kind of knowledge to persuade public opinion along the only secure path towards it, namely, the timely and liberal support of scientific investigation.

2.2.3 The MRC Survives the Squeeze

From the Report of the Medical Research Council for 1930–1, Cmd. 4008, 1931–2, pp. 128–9.

The MRC Report for 1930–1 concluded with reference to the effect of the 1930 economic squeeze on the Council's work.

The programme of work, of which a summary account has now been presented, has overtaxed the ordinary financial resources of the Council, and for reasons wholly independent of the results of the national financial crisis of last autumn. For the past four years there has been no increase in the grant-in-aid voted by Parliament for the work of the Council. In a service relatively so young as this, there must always be a rising expenditure as men in scientific or subordinate positions gain in normal increments of pay, and as permanent material equipments are being gradually built up. At the same time, many fresh and urgent demands have been made upon the Council's funds for work in some of the new and rapidly developing fields of medicine, while additional calls have been made upon them by other Government Departments for investigations of immediate practical relevance to the public service or the public health.

The Council referred last year to the heavy new responsibilities that have come to them in the work of determining and applying new standards of measurement for biological substances useful either as remedies or as foods. An international conference upon biological standards held in London last June under the League of Nations, to which detailed reference has been made at p. 51, made special calls upon the Council for preparatory work.

These various demands, though each of them may be taken as gratifying evidence of progress in well-founded work, have combined to overburden the ordinary resources of the Council. They thought it undesirable a year ago to ask for larger provision from Parliament to meet some of these additional burdens, and they met the adverse balance of their budget by making a special call upon money generously given them by private hands. Besides this emergency contribution to their work, the Council have received, as in many recent years, substantial contributions from public bodies and other voluntary sources to particular parts of their work that would otherwise have been quite beyond their reach within the limits of their Parliamentary grant.

With these circumstances in view it was naturally gratifying to the Council that the Committee on National Expenditure appointed last March found in their report of July no occasion to recommend any reduction in the expenditure of the Council or any change in their present responsibilities.

In consequence of the crises in national finance during the autumn, however, the Council have made increasing effort to diminish the total cost of their programme while preserving in every possible way its best value and utility. They proceeded at once to this task, and as one part of it they have called upon the whole of the workers in their service to submit, for the duration of the present emergency, to sacrifices in remuneration corresponding as nearly as possible with those inflicted in similar administrative and research services under the Crown.

2.2.4 The Growth of the DSIR through the 1920s

From the Report of the Committee of the Privy Council for Scientific and Industrial Research for 1930–1, Cmd. 3989, 1931–2, pp. 9–10.

The Annual Report of the DSIR for 1930–1 provided the opportunity to cast a reflective view over the evolution of the Department's organisation and growing responsibilities, and also to evaluate the progress the Department had made in encouraging cooperative research in industry.

It has been clear to each successive Advisory Council from the outset that the main endeavour must be to encourage industry to look on scientific research not as a last resource, but as an essential part of the business of production. With this end in view, they have made it their constant aim to define and emphasise the part which science properly applied can play in the country's industrial life, and to prove that the researches already carried out have provided technical assistance in the improvement of productive efficiency previously unobtainable. We believe that our predecessors' efforts have contributed substantially to increase the competitive power of British industry. Industry in general is proving itself increasingly anxious to take advantage of applied science. We have clear evidence that the assistance of the Department is becoming more and more in demand by industry, and we feel no doubt that it is now generally realised that the initial advantages which this country secured through her island position, her natural resources, and the technical skill of her workers are no longer sufficient in themselves to enable our manufacturers to withstand the organised and scientific rivalry of competing countries.

But while we are naturally encouraged by the more cooperative attitude displayed by industry towards scientific research, we are convinced that something further is called for by present-day needs. We gladly admit that, as the scientist and the industrialist by their increasing association come to appreciate the measure of their common task, so science is becoming more and more recognised as a profitable investment, and a useful ally in fighting industrial depression. But we feel that the attitude towards research is still too often one of hope rather than of faith; and that it has not yet become a matter of habit for all branches of industry to look to scientific research as one of the principal avenues of progress.

We are sure that science will not occupy its proper place with the industrialist until he is ready not merely to admit its possibilities and to accept its occasional assistance, but to incorporate it as part of his industrial practice; and we are looking to see him perform his part not only by contributing the means for carrying out investigations conducted in his interest, but by a readiness to cooperate with other firms for research purposes; by keeping track of the progress of research in matters of interest to his industry; by maintaining a competent scientific staff of his own; and by ensuring that the scientific outlook is represented, not only among the subordinate staff, but in the high places.

Unless this last point receives attention, there is a danger that research of the type here referred to will fail to yield the best results. These will be attained only if the direction of the researches is in the hands of persons who, besides being familiar with the scientific method and outlook, are alive to the economic and industrial aspects of processes which the researches may benefit or even create. Intimate contact between manufacturing practice and the work of the laboratory is essential if industrial research is to bear its proper fruit.

Scientific research has in the past made striking contributions to industrial progress and it will make them in the future. But the nation which will enjoy the benefits of science in the day-to-day progress of its industries is the nation which habitually applies scientific method and scientific knowledge; and it is that nation which will be able to seize the advantages of the more spectacular achievements of science in the industrial sphere. Scientific research cannot provide a ready-made solution of any of the present industrial difficulties; but it does point the road along which persevering effort may enable industry to find a way out of some of those difficulties. During the past twelve months of growing industrial depression our main preoccupation has therefore been to consider whether the Department can take any further steps than it has done already to ensure the application of science by industry. To this end we have made a comprehensive review of the arrangements for promoting contact with industry. We are satisfied that the steps already taken are in the right direction, and that the organisation which has been built up possesses the flexibility necessary to adapt itself to new needs as they may arise.

2.2.5 The Origins of the Agricultural Research Committee

From the Committee of Civil Research, Research Coordination Subcommittee, Third Report, April 1929, Public Record Office Cab. 58/106.

In addition to the historical survey of state-supported research, published as the first report of the Sub-Committee [cf. 4.3.8], two other (unpublished) reports were also produced dealing with more sensitive aspects of government policy. The third report dealt with agricultural research. The proliferation of agricultural research institutes and the spread of research between four or five different departments made coordination a pressing issue. A further factor was the need for agricultural research and development for the Empire. In its report, the Subcommittee, after reviewing current arrangements for agricultural research concluded that informal liaison was not sufficient, and that an Agricultural Research Council, on the lines of the other Research Councils, was necessary.

At their meeting held on the 18 October 1926 (Cabinet 53 (26), Conclusion 6), the Cabinet took note that, in accordance with the Second Report of the Select Committee of the House of Commons on Estimates (No. 119), recommending that more attention should be given to the coordination of Government research with a view to economy and efficiency, the Prime Minister had decided to ask the Committee of Civil Research to report under the following terms of reference:

'To consider the coordination of research work carried on by or under the Government, to report whether any further measures should be taken to prevent overlapping, to increase economy and efficiency, and to promote the application of the results obtained.'

3. In December 1927 we submitted our First and Second Reports. In our First Report (CR (C) 24) we endeavoured to draw a picture of the research organisation of Government in this country as it exists today. By direction of the Committee of Civil Research this Report was subsequently issued as a Stationery Office publication. In our Second Report (CR (II) 64) we submitted recommendations on various matters in which it seemed to us that some modification in existing practice would lead to better coordination of the scientific work of the Government. On that occasion we reserved for later consideration the question of the future organisation of agricultural research which forms the subject of our present Report.

6. Our survey of the existing arrangements which we have set out in our First Report (CR (C) 24) has satisfied us that a need exists for the establishment by the Government of some body charged with the duty of considering the present developments and needs of agricultural research not only in this country but also in the Empire in view of the connexions that have been and are being established between the research organisations in this country and corresponding organisations overseas.

7. Of equal importance and urgency is the establishment of an organisation dealing with agricultural research as a whole, which

can consult, and cooperate with, the scientific organisations already existing under Government in other branches of science. We refer in particular to the Department of Scientific and Industrial Research and the Medical Research Council, both of which have represented to us the importance of such a liaison from their respective points of view. The coordination of Government activity in these several fields is, indeed, one of the principal problems referred to us. It would be unfortunate if no steps were taken for several years to set up a body to carry out the double functions referred to in this and the preceding paragraph.

8. For the reasons given above it is impossible at present to establish a permanent body, but there may actually be advantages in setting up in the first instance a temporary rather than a permanent organisation. Beginning in a small way such a body would be able to learn valuable lessons from its own experience, and at the end of the transitional period it would be in a position to offer instructive suggestions for the consideration of whatever body was then charged with determining the form of permanent organisation to be adopted for agricultural research. If His Majesty's Government decide to establish a temporary body such as we suggest, we think it important that care should be taken to avoid any action which would imply the continuance of the organisation at the end of the transitional period on the lines first adopted, or its development in any one particular direction. Accordingly, we consider that the temporary body should not be attached either to the Development Commission or to the Lord President of the Council, or jointly to the Minister of Agriculture and Fisheries and the Secretary of State for Scotland. The suggestion which we put forward for consideration is that the Committee of Civil Research should appoint for this purpose an *ad hoc* Committee, the life of which should be definitely limited to the transitional period.

9. Such a Committee would of necessity have no funds at its disposal, and it would have to rely on its recommendations, backed as they would be with high scientific authority, being of sufficient weight to secure that effect would be given to them in appropriate cases by the Departments concerned and by the Empire Marketing Board. We recommend that this Agricultural Research Committee of the Committee of Civil Research should be entrusted with the following functions:

To consider and report on –

(*a*) Matters of importance to agricultural science to which in their opinion the attention of the Government shall be drawn;

(*b*) Questions involving cooperation between the English and Scottish Departments of Agriculture and other Government Departments in the United Kingdom, especially the Department of Scientific and Industrial Research and the Medical Research Council;

(*c*) Questions involving cooperation between the Departments of Agriculture or Agricultural Research Institutes in this country with corresponding Departments or Institutes overseas;

(*d*) Specific problems referred to them.

It is not our intention that the Departments concerned should necessarily refer to the Agricultural Research Committee questions affecting the normal development of, or the day-to-day research undertaken by, the existing Research Institutes. It will, however, be open to them to make such a reference in any case where they so desire, under sub-section (*d*) above.

12. The promotion of scientific research is now one of the major functions of the Ministers responsible to Parliament for Agriculture, and a large proportion of their votes are devoted to this subject. In fact, it is being increasingly recognised that in discoveries and application of science lie many of the main hopes of prosperity in the industry. It is therefore all important that Ministers themselves and their chief administrative officers should be regularly in contact with general scientific thought. This could be brought about in part by regular meetings of the proposed Agricultural Research Committee with the Parliamentary chiefs of the two departments. These meetings might take the form of the main Committee of Civil Research, consisting for this purpose of the Lord President of the Council, the Minister of Agriculture and Fisheries, and the Secretary of State for Scotland, and the members of the Agricultural Research Committee. It may also be desirable that the Secretary of State for Dominion Affairs and the Colonies should attend some of these meetings in view both of his responsibility for the Empire Marketing Board and of the growing importance of scientific work in connexion with agriculture in the

Empire overseas. The value of such meetings would depend on their affording opportunities to the scientific members of the Agricultural Research Committee for bringing to the personal notice of Ministers matters to which on scientific grounds they attach importance. At these meetings scientific questions could be brought up not only on the initiative of Ministers but of the Agricultural Research Committee. But this is only one method of improving the liaison between science, policy, and administration, and we feel that every opportunity should increasingly be taken both by Ministerial chiefs and their administrative staff to establish contacts with such a scientific group as is proposed.

13. If the Agricultural Research Committee is to prove an effective machine for the promotion of agricultural research it is important that it should be relatively small in size, so that each of its members may feel that he has a personal contribution to bring to its discussions. After careful consideration we recommend that the Committee should be constituted substantially on the model of the Medical Research Council, and should consist of eleven members, including the Chairman. In accordance with the normal practice of the Committee of Civil Research, the appointment of the Agricultural Research Committee would rest with the Lord President of the Council as Chairman of the Committee of Civil Research, after consultation with the Ministers primarily concerned, in this case the Minister of Agriculture and the Secretary of State for Scotland.

2.2.6 Creation of the Agricultural Research Council

From the Report of the Agricultural Research Council for the period July 1931 to 30 September 1933, Cmd. 4718, 1933-4, pp. 3-7.

In 1931 the arrangements proposed by the Research Coordination Subcommittee of the Committee of Civil Research [2.2.5] were formalised by the creation of the Agricultural Research Council, to complete the triumvirate of research councils. In this first report the Council described its administrative and financial arrangements and projects some of its future activities.

The initiation of a new body such as the Agricultural Research Council inevitably involves administrative problems requiring careful and unhurried consideration, especially when the new body necessarily stands in close relation to other pre-existing bodies. It was only after prolonged examination of those problems by Committees appointed by the Cabinet that the functions of the Council were defined.

Firstly, the Agricultural Research Council was designed to complete the Scientific organisation for the supervision of subsidised research, other sections of which were already managed by the Medical Research Council and the Department of Scientific and Industrial Research. . . .

Secondly, while the Medical Research Council and the Department of Scientific and Industrial Research have themselves built up and now control the Research Institutes and isolated researches for which they are responsible, when the Agricultural Research Council was founded there was already in being an extensive organisation for agricultural research, initiated under a scheme framed by the Development Commissioners in 1911, and since then extended in collaboration with the two Departments of Agriculture, who are responsible for the administration of the large Development Fund advances provided annually for the maintenance of this organisation.

In the Lord President's Provisional Note on Administrative Arrangements, it is laid down that, while the general relationship of the Agricultural Departments with Agricultural Research Institutes, University Departments of Agriculture and Agricultural Colleges is to continue unchanged, the Agricultural Research Council is to give criticism and advice on research grants and programmes, though the expenditure for research services hitherto administered by the Agricultural Departments and borne on their Votes is so to remain, and Parliamentary responsibility for agricultural research remains with the Ministers responsible for agriculture. On the other hand, the administrative expenses of the Agricultural Research Council are borne on the Vote for Scientific Investigation, etc. (Class IV, 9), under the Parliamentary responsibility of the Lord President of the Council, and the special research grant (£5,000 in 1931 and £6,000 in 1933) drawn from the Development Fund, a grant which also now appears on that Vote, is to be spent on their own authority by the Agricultural Research Council under the general direction of the Committee of Privy Council, the Departments of Agriculture being consulted on proposals affecting Institutions for which the Departments are responsible. It is further laid down in the Note that the Council shall act as the scientific advisers of the Development Commissioners. The dividing line between expenditure financed direct by the Agricultural Departments, and expenditure financed by these Departments by means of grants from the Development Fund, is to be retained.

It is clear that the relations of the Agricultural Research Council with these bodies – on the one side with the Medical Research Council and the Department of Scientific and Industrial Research, and on the other with the Agricultural Departments and the Development Commission – make the administration of its work both complicated and difficult. Indeed, it could only be carried on successfully with the general good will of all concerned. . . .

The Report of the Research Council to the Committee of Privy Council, appended hereto, shows that the Agricultural Research Council began its work by initiating, with the encouragement of all the Departments concerned, a comprehensive survey of the agricultural research now being carried on in Great Britain, especially in those Institutes partly or wholly supported by public funds. In order to facilitate the work, six standing Committees of the Council were formed . . .

During the time now under review, visits were paid to twenty-two Research Institutes or other research centres either by one of these Committees, or by two or more of them conjoined, sometimes followed by visits of special *ad hoc* Subcommittees, including in their membership eminent experts from outside the Council.

Some of these visits were arranged in the normal course of the Council's general survey of agricultural research, but others were specially undertaken in order to answer quickly appeals for advice made to the Council by the Ministry of Agriculture, the Scottish Department of Agriculture or the Development Commission. Special reports have been rendered in answer to such questions.

Among these special reports we may mention firstly one containing the results of a careful inquiry on the past history and future prospects of the research on Foot-and-Mouth Disease which, for the last ten years, has been carried on under the direction of a Committee of the Ministry of Agriculture and Fisheries. The Agricultural Research Council formed a special expert Committee to make this inquiry. As will be seen below, the Committee reported that it was convinced that excellent work, resulting in definite advances in knowledge, had been done. While it could not be said with certainty that any preventative measures for this disease, such as an effective vaccine, were likely to be discovered immediately, such measures, based on the results already reached, were certainly a possibility of the future. The Committee also made recommendations for the future scientific administration of the Foot-and-Mouth Disease Research Station.

Again, at an early stage of its general survey, the Agricultural Research Council came to the conclusion that at the present time the most urgent subject of inquiry for the economic well-being of the British farmer was that of animal disease. This conclusion was strengthened when Your Majesty's Government foreshadowed an agricultural policy which involved an increase in the animal population of the country, and a consequent additional risk of the spread of certain infectious, contagious, or otherwise communicable diseases.

To meet the situation, the Agricultural Research Council, after consultation with veterinary experts, resolved to institute an immediate inquiry into the present state of knowledge regarding the most important and most dangerous of such diseases and the work in progress upon them, and formed seven special Committees, most of them consisting mainly of experts from outside the Council, to undertake the work.

Reports from these Committees are now coming in. Some of them contain recommendations for immediate research, and in certain cases the Agricultural Research Council has either itself allocated funds or made arrangements for the necessary provision through a Department of Agriculture, and authorised the appropriate Committee to pass from advisory functions to those of general coordination. The Committee of Your Majesty's Privy Council hopes that thus the Agricultural Research Council may do its share in making possible the immediate agricultural development which Your Majesty's Government has in mind.

2.2.7 Criticisms of the DSIR

From J. D. Bernal, 'DSIR Annual Reports for 1933–34 and 1934–5', *The Scientific Worker*, March 1936, pp. 40–2.

In this critical review of the work of the DSIR over the period 1933 to 1935 Bernal attacked the piecemeal approach and trivial pre-occupations which, he said, characterised its work. In the conclusion to the article he proposed an Industrial Research Department which would link industrial, health and agricultural research.

Taken by themselves the results of the activities of the Department indicate a steady though somewhat pedestrian tackling of industrial problems, but in regard to the needs of industry or the possibilities of scientific research they represent only a small fraction of what ought to be and could be done. Nevertheless it is possible to draw from a study of the methods adopted by the Department in these fields some indication of how the larger task of the application of science to human welfare should be attempted.

The function of science in relation to production is two-fold: to increase the effectiveness of current methods and to search for radically new or improved methods of production. There are in the present system of production almost insuperable obstacles to the latter course. For the Government to take a positive line in research would imply entering into production and this is not welcome either to competitive or monopolistic industry. Government research cannot in practice be carried on without the willing cooperation of industrial firms and consequently it is obliged to restrict its activities to the improvement of present methods of production. The function of applied science at the moment is simply that of reducing cost, and this, except in the peculiar case of the Safety in Mines Research, without any direct regard to improving the conditions of the workers engaged in production.

Most industries have grown up in a technical rather than scientific way (the noted exceptions are, of course, the electrical and parts of the chemical industry). The first function of science is therefore to study the industry, in the same way as the biologist studies the animal or ecological process in which he is interested. This involves the development of new types of apparatus adapted to the particular industry. An example of the value of such work is the instrument developed in the wool industry which enables a certain type of damage to be detected in the course of twenty-four hours instead of allowing three weeks for it to develop naturally. The scientific study of industry, however, requires not only to be intensive in this sense, but extensive. It is no use examining each process in detail without relating it to the processes of the whole of the industry. Scientific research is required just as much for the design of an economically-run plant as it is for the design of any particular material or process.

Once the nature of the industry has begun to be understood scientifically, it is possible not only to measure and standardise, but to introduce economies and to reduce wastes. Most of the valuable routine work of the Department has been of this kind. The introduction of new processes – which it rarely initiates – leads to unexpected flaws in operation, and these require to be corrected by scientific research. But it is plain that this is an extremely roundabout process. The planning of an effective industry and the correction of the mistakes in its running should be undertaken together. But in a profit-making society, the two tend to get separated, and the Government Departments tend to have to do for the most part with the pathology of industry, and not with its health and growth. Profit taking makes it impossible to have a rational scientific approach to industrial problems. We do not distinguish at all between the continually useful type of profit – that introduced by improved methods – and the harmful form which consists in monopolising methods or even in restricting them to certain firms who thereby acquire purely differential advantage. If the whole of industrial production were opened to scientific improvement, a great deal of the remedial work for which scientific research is now used would be found to be unnecessary. Difficult processes involving much waste and complication would be rapidly superseded.

But it is not sufficient to consider an industry either intensively or extensively; every industry stands in relation with producers of raw materials, and with consumers, whether intermediate or final. Every research institute requires a close, organised connexion with corresponding institutes in the raw material field, i.e. those of agriculture, fisheries, mining and quarrying, and correspondingly with the consumer research bodies, either those of other industries or the, as yet non-existent, individual consumers' research associations. Why have we no such consumers' research association? Simply because consumption in itself is not a profitable enterprise, though it is the enterprise from which all profits are derived. Actually, if one-tenth of the funds now devoted to merely competitive and wasteful advertising were devoted to consumers' research, not only would the people at large receive more real and conscious satisfaction, but there would result a stimulus to industry which in the total would probably be greater than that derived from the forced selling of the advertising system.

The whole of industrial scientific research requires to be a closely linked system, not only of the sort already described, but with corresponding horizontal links for industries of the similar types – as, for example, between all metal industries, and those of coal and power, or between the chemical industries and the food industries. Scientific research, particularly industrial scientific research, has a value which depends far more on its relations to other work than on anything intrinsic. Given a certain supply, albeit inadequate, of research workers and funds, it is not sufficient to say that these resources are all being used to produce good research results. The research requires to be organised so that only that which most needs to be done is attempted, and this necessarily involves a close collaboration with economics at all stages of industrial research. Up till now, economics only enters research in the rather cumbersome and thoroughly unsatisfactory method of dependence on the conscious research needs of individual firms.

It may be that it is radically impossible in a system of industrial competition to have a rationally organised research system. But that is not to say that it would be impossible to complete the present type of research system and to extend its activities considerably. The Department, at any rate, has recognised these possibilities. On page 12 of the 1934–5 report it states:

'We have referred to the encouraging evidence of the change which is being steadily brought about, largely by the wholehearted efforts of enlightened leaders in our great industries. But that much remains to be done is only too evident, and it is important to remember that we cannot afford to lose any time. This is made clear from a report of an inquiry made by one of the Research Boards of the Department at our request. We had invited the Board, which contains representatives of industry of undoubted authority, to review the conditions in the particular industries with which it was concerned, and to say whether sufficient research was being undertaken in the country for their benefit; if not, in what directions further investigation is called for, and what further steps are needed to impress on these industries the need for better application of existing knowledge and for further research work. We were much impressed by one conclusion which emerged clearly: that, compared with some of their industrial rivals overseas, the scientific outlook of these industries still leaves much to be desired. Neither in the directorates nor among the technical and executive staffs is sufficient weight yet given to scientific attainment and experience; and until a radical change has taken place in this respect, the position is bound to be that the industries as a whole will remain unable to obtain the full benefit of the results of scientific investigation.

'The truth appears to be that many of our industries are still paying a price for the remarkable era of prosperity which followed the industrial revolution in this country and lasted till the Great War. An example is afforded by the extent to which manufacturers of machinery have allowed their foreign competitors to establish commanding positions in the world market for new types of machines by the systematic application of scientific investigation and invention.'

The present system of detailed attacks on relatively trivial problems in industry gives no inspiration either to the industrialist or the general public, and does not help to justify to anything like the degree that it is intrinsically justifiable the value of scientific research. It is only by coming into the open with bold and comprehensive schemes that there is any hope that adequate support for productive scientific research will be found. What we require is the building up of an Industrial Research Department which covered not half but all industries in the country and was closely linked to similarly developed agricultural and health research departments. Some means should be found for supplying it with a secure and steadily increasing budget so as to enable long-range investigation to be confidently undertaken in attracting to the services of industrial research an increasing number of the best scientific abilities. The services such an organisation would give to industry and the public would soon be recognised to repay the extra expenditure entailed. It is only in taking part in such a scheme that the scientific worker could be effective in his own work and feel a factor in social progress.

2.3 COORDINATION, SCIENTIFIC ADVICE AND THE EMPIRE

2.3.1 First Attempts at Coordination

> From the Report of the Committee of the Privy Council for Scientific and Industrial Research for 1919–1920, Cmd. 905, 1920, pp. 25–6.

The first attempt to create some systematic machinery to prevent the wasteful duplication of research was the setting up of the coordinating boards in 1921. The decision followed from the recommendations of the Committee on the Coordination of Scientific Research in Government Departments which had been created in 1920 by Lord Balfour, in response to Treasury anxiety about the research budget. The Committee's recommendations included a system of informal yet regular meetings between representatives of the civil research departments, and a formal network of coordinating boards to ensure more efficient communication between the research departments of the fighting services.

At the beginning of this year the Government determined that steps should at once be taken so to organise the scientific work that was needed for the fighting services as to avoid unnecessary overlapping, to secure the utmost economy of personnel and equipment, to facilitate the interchange of scientific knowledge and experience between all the departments concerned, and to provide a single direction and financial control for all work of a fundamental nature of civilian as well as military interest. They decided that the best means of achieving this would be to establish a series of boards upon which each of the fighting services, and any civil departments of State engaged to a material extent in research connected with the work of the boards, would be represented by responsible technical officers; and that with these there should be associated individual men of science. The boards were to be attached to the Department of Scientific and Industrial Research and would be appointed by the Lord President on the advice of his Advisory Council. They would, under the Department, direct any research of a fundamental nature that might be required, and any investigations having a civilian as well as a military interest, while the cost of such researches would be borne upon the Vote of the Department, though the work might be done in the establishments and by officers in the employ of some other Ministry. The Department would also be responsible for securing contributions from industry in suitable cases, and it would arrange for the cooperation of industrial research associations where this was appropriate.

The Government directed that in the first instance three boards should be established, one for chemistry, one for physics, and the third for engineering, and that these with the existing Radio Research Board should form the nucleus of the scheme. It was recognised that it would be necessary to appoint subcommittees and probably joint committees of the several boards to deal with particular subjects.

The Radio Research Board had been established somewhat earlier by Your Lordships on our recommendation as the result of representations made to the Department by the Imperial Communications Committee of the Cabinet. We summoned a conference of all the departments concerned to consider the proposal, and with their unanimous concurrence we advised Your Lordships to set up the Board with the following terms of reference:

(a) To provide for interchange of information between the various Government technical establishments concerning the special work which they undertake and the results achieved, so as to prevent duplication of work; though the Board should have no executive function as regards the work of these establishments.

(b) To arrange for the communication of such information to interested persons outside the Government service when this can be done without detriment to the public interest.

(c) In the case of researches not otherwise adequately provided for, to make the necessary arrangements to meet the requirements of Government Departments and others.

The same terms of reference have been given to the three new boards and their personnel, with that of the Radio Research Board, will be found in Appendix I.

We realise the great importance and difficulty of the new responsibilities thus placed in our hands, but we believe the new arrangements will conduce greatly to economy of effort and of money, and to increased rapidity of progress if all concerned will help forward this interesting experiment in the cooperative conduct of research. If firms competing with each other for existence can combine, as they have done, for their common benefit, it ought not to be more difficult for the members of a national service to do so, merely because they are attached to different departments of the Government.

2.3.2 Tizard and the Coordinating Boards

From Sir Frank Heath, letter to Henry Tizard, 5 May 1920, reprinted as an appendix in *Tizard*, by Ronald Clark (Methuen, London, 1965).*

Heath gives the further details of the post of Assistant Secretary to the Coordinating Boards after having offered the job to Tizard.

Dear Tizard . . . you begin with a general statement of what you understand the policy to be so far as research for the fighting services is concerned and I think it accurately represents the position, though I should like to add one or two guarding statements.

You say these Committees 'will examine the proposals critically, will endeavour to coordinate the research work of the three departments and will provide such work as *they* decide should be carried out'. That is the intention, but obviously it will be necessary to make sure that the decision of the Board for undertaking the special piece of work themselves could safely be acted upon. It could not, at any rate in the early days, without our satisfying ourselves that their decision would not be opposed by any of the Departments represented upon the Board. It would fall to the Assistant Secretary in charge to satisfy himself that the supreme authorities in the fighting Departments or other Departments represented, were in agreement with the proposals put forward before action was actually taken. This, you will realise, would be one of the important duties of the high official in question.

In a later sentence you go on to say that 'investigations will be carried out either at existing Government experimental establishments or at the Universities, private firms, etc. whichever is most suitable'. I agree that we cannot safely exclude the possibility of research for the fighting services by private firms, but the actual

* Where the date is incorrectly given as 1926.

arrangements are not easy, and will need to be carefully considered in the light of the general attitude of the Department to private firms – a delicate problem, which is not by any means solved at the present moment.

I now come to the definite points which you put to me:

(1) The Assistant Secretary will be responsible for seeing that the decisions of the Committee, so far as they fall within the approved programme of research, are put into action. Our idea is that each of the more important Boards or Committees will have a Technical Officer of the Department attached to them, whose duty it will be to prepare the technical material for the Committee under the general directions of the Assistant Secretary, and also to act under him if he thinks desirable in visiting laboratories and the like. . . .

(2) It is not contemplated that the Assistant Secretary should be a member of each Board, but he will always of course be able to attend when he thinks it desirable, and it will sometimes be important that he should be present in order to explain the policy of the Department and keep the Boards for which he is responsible in line with each other. The Committees or the Boards will be kept as small as is consistent with a due representation upon them of each Government Department directly concerned and of the inclusion amongst them of independent men of science appointed by the Department through the Advisory Council.

(3) We contemplate that the Assistant Secretary and the Boards will have direct access to all experimental establishments in which any work is proceeding under their direct scientific control but, as you say, the administration of these Stations will be unchanged, and the control on our side will be confined to the work for which one of our Boards is responsible and will be scientific rather than administrative. . . .

Your reference to the research work at Farnborough which is of importance to industry in general, touches upon a part of the field where there is likely to be difficulty, because, as you know, before the coordination policy of the Government was adopted, the cabinet had decided that the Air Ministry was to remain responsible for all research work dealing with civil aviation, except in so far as it might be undertaken by a Research Association under this Department or as part of the general policy of the Department for aiding research in pure science.

If you will look at Page 11 of *The Times* for today you will notice that there are three clauses defining the respective spheres of the Air Ministry and this Department. Clause C contemplates the establishment of permanent machinery for general coordination of Government research, and so far as the Air Ministry is concerned this general coordination has to be worked out. The devising of a workable arrangement will fall largely upon the new Assistant Secretary, and that is one reason, I may say, why I should like to see you in this Office; but if we may take as an example some other research establishment for the fighting services, such as Woolwich, where the particular complication that exists in the case of the Air Ministry is absent, your assumption that the whole of their scientific investigations should be submitted to the appropriate Board is right. I have no doubt that we shall find a certain unwillingness to place all cards on the table – indeed there is some evidence of it already – but with patience and tact I believe this can be squared, for we have the Treasury firmly behind us in this policy, and if any Department is unwilling to play the game it will soon find that its means are cut off at the source. It is quite true that in practice research work is likely to be initiated by Officers of the different Departments, and the money difficulties are insufficient to ensure that the Department concerned will keep the Boards informed, but, as you rightly say, it is an important part of the work of the new Assistant Secretary to see as tactfully as he can that everything possible is referred to the coordinating Boards.

(4) As you will have gathered from what I have already said, the Aeronautical Research Council is just the most difficult part of the whole problem. The trouble is, it does not fit in to the new policy of the Government, and it ought, as you said, to be a committee of this Department; but we shall have to be slow in this matter, and begin by a working agreement which may lead to a tidy organisation later. The new Admiralty experimental Station at Teddington ought to be much easier to deal with because that Station will be confined to doing work of a secret nature for purely Admiralty purposes. All work of wider interest and the preliminary investigations upon which these secret applications will be built up is to be done at the National Physical Laboratory buildings, and I have no doubt whatever that the amount of work done in their own Laboratory will form a very small proportion of the total amount which they will wish to see done.

Turning next to your personal difficulties, I think you must recognise that the post is primarily administrative, but it is administration of a kind which deals throughout with scientific work and research. It would be almost impossible for anyone to fill the office effectively who was not a man of scientific standing. The amount of scientific initiative which the new Assistant Secretary can undertake will depend almost entirely upon himself. I am quite certain that he will not be discouraged, and if it were not exerted it would be at the least unfortunate. On the other hand, the initiation must, I think, be personal rather than official, just because of the doubts which might arise in the minds of other Departments if an officer in the powerful executive position of the Assistant Secretary had also the means of initiating research on behalf of this Department in his official capacity.

This sort of illogical position is, as you know, very common in this country. If scientific initiation were definitely stated as one of the Assistant Secretary's functions, it might lead to suspicion, while if he is the right man he will have ample opportunity in fact of exercising it. It would be absolutely essential that the Assistant Secretary should, as you say, 'Try to get and keep the confidence of the scientific, as well as the military, world', and to this end his connexion with Scientific Societies would be vital. I also agree that the more direct contact you could have with the investigations for which the Department and the Boards would be responsible the better, so that, as we agree, your work would by no means be confined to the Office, though in the early days I am afraid you would find it necessary to be on the spot a good deal.

2.3.3 Scientific Advice and Imperial Development: the Formation of the Committee of Civil Research

From Sir Thomas Middleton, Sir Frank Heath and Sir Walter Fletcher, A Memorandum on the Formation of an Imperial Development Council, February 1925, Balfour Papers (British Museum, Add. Mss. 49753), ff. 16–32.

In April 1925 Lord Balfour became Lord President of the Council for the second time. He inherited a scheme for a Committee of Economic Enquiry which had derived originally from the fertile brain of Lord Haldane, and had been taken up in part by Ramsay MacDonald before he lost the election of October 1924. Balfour, in the spring of 1925 seized the opportunity to create a Cabinet Committee of scientific experts, analogous in function to the Committee of Imperial Defence, to study scientific and economic problems relevant both to the United Kingdom and to the Empire. The following extract is taken from a memorandum prepared by three senior civil servants, on the need to exploit scientific advice for imperial development. Their call for action was, in many ways, answered by the ambitious research programme of the Committee of Civil Research which investigated problems ranging from locust control to nutritional deficiency in the Empire, and from the coordination of government research to the steel industry at home.

The Government have under consideration measures for developing productivity throughout the Empire. The Prime Minister recently stated in the House of Commons that a sum of £1,000,000 a year would be available in the immediate future as a means of 'improving the methods of preparing for market and marketing within the United Kingdom the food products of the overseas parts of the Empire with a view to increasing the consumption of such products in the UK in preference to imports from foreign countries and to promote the interests both of producers and consumers'. In *The Times* 30 January Mr Amery is reported as having said,

The development of our Empire was not only a great political and administrative problem. It was also, and no less a scientific problem – a problem of applying scientific research to the practical task of making the most of our immense resources.

The need for enlisting scientific aid in developing their resources has long been recognised by many of our Dominions and Colonies and substantial results have already been secured. In India much success has been met with by plant breeding, especially in the creation and introduction of valuable new wheats. ... In South Africa, valuable discussions have followed the provision of a laboratory for investigating the diseases of animals. The exhibition at Wembley last summer supplied much evidence of the progress made by scientific workers in the Dominions and Colonies. ...
In this country we were later in attempting to organise scientific aid for the solution of our domestic problems than were several of

the Dominions. But within the past ten years great advances have been made, and we are now in possession of highly developed systems for the systematic investigation of the problems of primary producers, of our manufacturing industries and of disease in man and domestic animals.

In these developments at home and overseas much experience has been gained of the conditions which are essential for success in bringing science to bear on the development of our resources; but there has been no systematic attempt at exchange of experience or at coordination of the similar work in progress in the Mother Country and in other parts of the Empire, still less is there any machinery by which initiative can be taken in new directions. Such contact as exists between scientific men at home and overseas is maintained by Universities, scientific societies and the circulation of scientific journals. Various Government departments at home have vital interests in these matters, but neither in theory nor in practice is there any continuous interchange of experience and effective coordination of work. Nor is there in practice any effective arrangement for what may be called the 'scientific staff work' for meeting imperial needs by which the work of the Government departments at home may be correlated with that of the self-governing dominions and future needs integrally anticipated.

'The economic development of the Empire is a scientific problem.' At bottom it is a problem of applying physiological and medical science to men so as to make economic development possible: medical science, including gaining and using the knowledge necessary for the prevention of disease, the treatment of disease, the proper management of healthy human life under varied conditions of labour and climate, the right guidance of particular economic schemes, and the preparations to be made for effective naval and military defence. At the next stage the economic development of the Empire is a problem of applying physiological and genetic science to plants and animals, of applying chemistry, physics and engineering to mineral resources, of applying the same sciences to the industries that grow out of these primary means of winning the fruits of the earth, and finally, it is a problem of communications. What is needed is a single ordered campaign of scientific effort on all these interlocked lines of development.

2.3.4 The Committee of Civil Research and the Coordination of Research

From *Nature*, 7 July 1928, vol. 122, pp. 1–3.

One of the most important investigations of the Committee of Civil Research was that on the coordination of research in government departments, through its Research Coordination Subcommittee, in response to renewed Treasury anxiety about the evolution of the research services of government and the fears that some were redundant, some unnecessary and many duplicating work that was more efficiently done elsewhere. The first Report of the Subcommittee published in 1928, a general survey of the historical evolution of state-supported science, was a worthy historical document but one which was hardly a critical review of the current situation [4.3.1]. The report brought derisory reviews from *Nature* which argued that, given the opportunity, the Subcommittee should have produced something more in the nature of a policy document.

In consequence of a recommendation of the Select Committee of the House of Commons on Estimates, the Prime Minister, in October 1926, constituted a Subcommittee of the Committee of Civil Research 'to consider the coordination of research work carried on by or under Government, to report whether any further measures be taken to prevent overlapping, to increase economy and efficiency, and to promote the application of the results obtained'. Mr Ormsby-Gore was appointed chairman of the Subcommittee, Major Walter Elliot was the other parliamentarian, the other members being permanent officials of the various departments concerned in the inquiry. The inquiry was spread over a year, the Subcommittee reported on 14 December 1927, and the results of its deliberations were made public on 14 June last.

With such terms of reference, it might reasonably have been expected that the Subcommittee would consider it its duty to present a critical survey of the present organisation of the State-maintained or assisted research services of Great Britain, to present broad details of the expenditure involved, to express its own views on the relative merits of the varying types of organisation which came under its purview, and to have given some indication of the directions in which the results of research might be applied with advantage to the community, and the new researches that were necessary. In these respects the report of the Subcommittee is disappointing. It is critical of the attitude of indifference to research of former generations, but it is difficult to detect in its complacent generalisations on the present organisation of research whether it considers any changes desirable or that it is satisfied that the existing machinery works to the best advantage.

The report is evasive and apologetic in turn. It gives the impression that expenditure on research has to be defended against possible attacks by the ignorant, rather than justified to those who understand the real purpose of research and whose aim it is, in the interests of scientific research generally, to exact the greatest degree of efficiency from the instruments by which it is conducted.

On only one subject of importance has the Subcommittee expressed definite views. These occur in the last three paragraphs of the report and deal with the publication of scientific knowledge. The Subcommittee rightly considers that the present variety of means adopted by different Government departments for the publication of results of scientific value is not an advantage. Each department has been a guide to itself in the matter, and sufficient account has not been taken either of the need for coordination and uniformity of presentation of results obtained by men of science in the Government service, or of the importance of regarding their contributions to knowledge as contributions to the common stock, inseparable from those of scientific workers outside the Government service. 'It is,' says the Subcommittee, 'incumbent on the Government to avoid adding to the mass of publications that must be searched by scientific workers if there already exist adequate means for the purpose in the scientific world.' It considers that the most effective publicity for results is obtained by means of the *Proceedings* and *Transactions* of the various learned societies and technical journals, hitherto 'undertaken at the charge of individual workers banded together for the purpose'. It therefore envisages the possibility of more extended use being made by Government departments of these agencies and of direct State contributions towards the cost of such publications. Evidently it considers that the increase, made in 1925, of the Treasury grant to the Royal Society in aid of publications, has been thoroughly justified, and it is permissible to assume that an application for a further increase would be received favourably.

This is the only bright spot in an otherwise dull summary of the methods by which the State fosters research, either in Government laboratories staffed by professional Civil Servants or in State-maintained or assisted research institutions. It is conceded that such a summary will serve a useful purpose; it might, for example, stimulate more parliamentarians to take an active interest in a matter of vital importance to the nation by enabling them to appreciate the influence of scientific research on our social and economic life, but it was scarcely necessary to have called together so eminent a body to compile what appears to be a digest of various departmental memoranda, the only excursion into matters of policy being that noted above.

2.3.5 Epitaph for the Coordinating Boards

From the Committee of Civil Research, Research Coordination Subcommittee, Second Report, December 1927, Public Record Office Cab. 17/121.

The second unpublished report from the Sub-Committee [cf. 2.2.5] dealt with the problems of the coordination of civil and military research. This extract is critical of the work of the Coordinating Research Boards and recommended their abolition in favour of more informal liaison between the fighting services. Clearly the Boards had failed and new machinery was needed.

21. The Select Committee on Estimates drew particular attention in their Report to the need for further coordination of the Research Organisations of the Fighting Services, and we therefore devote a special section to this subject. The existing system of Coordinating Research Boards established by the Department of Scientific and Industrial Research is described in our First Report.

22. These Boards have no executive functions, and their responsibilities as regards coordination are, in practice, confined to the interchange of information between the various Government technical establishments. They have endeavoured to carry out this duty by asking annually for particulars of the research programmes of the Fighting Services and arranging for a general discussion of these programmes at meetings at which technical representatives of the Services are present. The Service Departments have supplied the Department of Scientific and Industrial Research with a

great deal of information to enable this to be done, but, naturally, they have not been able to supply it with much information on secret researches. Decisive action has seldom, if ever, resulted from the discussions of the Boards. That the discussion of these common programmes has, however, done some good there is little doubt. The Boards have rendered a useful service in furnishing an informal meeting-ground for the scientific officers concerned. Their procedure is, however, laborious, and it appears doubtful to us whether the results secured are really worth the time and labour involved. The Boards cannot compel any Service Department or any section of any Service Department to stop research on the grounds that some other Department is doing similar work. All they can do is to draw attention to the possibility of overlapping. Their position is, moreover, somewhat anomalous, as they possess a certain degree of responsibility, but have no power of enforcing their views against those of the Service Departments. The grant of such powers would, in effect, involve the transfer of responsibility for the direction of Service research from the military staffs of the three Services to a civilian Department. Such a transfer would involve grave difficulties in regard to the Ministerial responsibility of the Service Ministers. On the uninterrupted development and application of research much of the efficiency of the Fighting Services must necessarily depend, and separation of responsibility for such an essential service would be wholly inconsistent with responsibilities for national defence imposed on the Board of Admiralty and on the Army and Air Councils. Even if these difficulties could be overcome – and we are of opinion that they could not – the transfer of responsibilities for research to a civilian Department would defeat the very ends which it was designed to serve. A research organisation can only be expected to be effective insofar as it is closely associated with the practical needs of the Service for which it is working. The effectiveness of Service research would, in our view, be materially impaired if the present intimate relations between the research organisations and their respective Services were disturbed. Such a result would inevitably follow the transfer to a civilian Department of responsibility for the conduct of research. No civilian Department would be able to acquire at second hand the intimate knowledge of Service requirements possessed by the Fighting Department. Without such knowledge it would never succeed in giving to research the direction required by the peculiar conditions of the three Services, and its work would tend to become more and more unreal in character. Such results as it might obtain would, moreover, often fail to secure the acceptance of the Services.

23. The conclusions which we have drawn from our review of the organisation set up in 1919 and of the subsequent developments in the organisation of the Fighting Services for Research make it impossible for us to endorse the following recommendation contained in the Second Report of the Select Committee on Estimates (No. 119):

> 'Your Committee are also convinced that more attention could be given to prevent overlapping of the work of Scientific and Industrial Research undertaken by the Admiralty, the Air Ministry, and by the Department of Scientific and Industrial Research. It is felt that no opportunity should be lost to bring this work, so far as possible and consistent with economy and efficiency, under the control of one Department.'

24. We believe that the most hopeful line of progress lies in the direction of improving the means at the disposal of the Fighting Services for effective cooperation rather than by a formal superstructure of coordinating machinery. Substantial progress has been made in this direction by the reorganisation at the Admiralty and the Air Ministry, to which we have referred, and the position will, we believe, be further improved if our proposals (paragraphs 17 to 20) in regard to the War Office are adopted. These developments have robbed the Coordinating Research Boards for Chemistry, Physics and Engineering of their former usefulness, and we recommend that they should be dissolved.

25. Effective coordination between the Services in future can, it appears to us, be secured by the appointment, from time to time, of special *ad hoc* Committees for the consideration of particular problems. These *ad hoc* Committees should, we recommend, take the form either of Inter-Departmental Committees, or, in the case of inquiries with a wider scope, of Subcommittees of the Committee of Civil Research. However constituted, they should contain representatives not only of the Fighting Services, but also of the Department of Scientific and Industrial Research, and, where desirable, of the two Departments of Agriculture and of the Medical Research Council. It should, in our view, be open to any

of the Fighting Services or to the Department of Scientific and Industrial Research or other Departments interested to suggest the appointment of *ad hoc* inquiries of this character. Discussions of these and similar problems would be facilitated by periodical meetings between the Directors of Scientific Research at the Admiralty and the Air Ministry, the proposed Scientific Adviser at the War Office and the Secretary of the Department of Scientific and Industrial Research, and we recommend that steps should be taken to establish meetings of this kind on the dissolution of the existing Boards; and that each officer should report annually to his Department on the results of these meetings.

2.3.6 Proposals for the Coordination of Scientific Research and Advice in Wartime

> From Sir Henry Tizard, FRS, Proposals for a Central Scientific Committee under the Minister for Coordination of Defence, 21 July 1938. Public Record Office Cab. 27/711.

> Towards the end of the 1930s the growing threat of German aggression stimulated several suggestions for the preparation and organisation of scientific research and advice in the event of hostilities. Both the government and the scientists bore the lessons of 1914 clearly in mind. The following extract is from a memorandum submitted to the Cabinet in July 1938 by Tizard for a small Central Scientific Committee to coordinate scientific research in the armed services.

1. I sent last month to Sir Kingsley Wood a few notes about the working of the Air Ministry Research Committees. I also threw out the suggestion that it might be wise now to appoint a small Central Scientific Committee under the Minister for Coordination of Defence. Sir Kingsley Wood tells me that you would like me to develop this idea a little.

3. My proposal is strictly for a Research Coordination Committee. There has now been appointed a Director of Scientific Research at the War Office. It seems to me that it would be of the utmost value on general grounds to bring the Directors of Scientific Research in the three Defence Departments together at reasonable intervals, say monthly or quarterly, and to assist them with the

collaboration of selected outside scientists. I do not believe in trying to cover the whole field of the application of science to defence problems. I would not, for instance, include biology with the physical sciences, as was originally suggested. I am merely thinking of the application of the physical sciences to the equipment and methods of defence.

4. Much of my experience during the last three years goes to show how necessary it is to get complete collaboration between the Scientific Departments of the three Ministeries [*sic*]. There has always been friendly collaboration between the Directors of Research at the Air Ministry and the Admiralty, and between them and the various individuals responsible for research and experiment in the War Office. But something more than casual friendly collaboration is needed. If we are striving to apply the latest advances of science to the problems of defence, it will only be rarely that any application affects one Service only. For instance, the scheme, produced for Air Ministry purposes, has most important applications to the requirements of the Admiralty and War Office.

9. To sum up, my proposal is that a Research Coordination Committee should be set up under the Minister for the Coordination of Defence, and that its personnel should include the three Directors of Research of the three Service Departments and the Secretary of the Department of Scientific and Industrial Research, with not more than six independent scientific members; and that its terms of reference should be:

(*a*) To initiate and to advise on proposals for the application of new scientific methods to problems of defence.

(*b*) To advise on any scientific problems specifically referred to them by the Service Departments and by the Committee of Imperial Defence.

(*c*) To advise on the selection of individuals to assist the Service Departments by scientific advice on particular problems of defence.

(*d*) Generally to keep in review the organisation of the scientific research of the Service Departments.

10. The first and most important of these terms of reference requires suitable financial backing. What I suggest is that such a committee, if formed, should be able on its own authority to direct

the expenditure of money on new inventions or methods, such expenditure being under the administration of one of the Directors of Scientific Research. In other words, in all the initial stages the committee would have the responsibility of expenditure, but one of the Departments would be in executive charge. It seems to me that for a comparatively small commitment it should then be possible to avoid any delay in the initial stages of important work. If at the end of the preliminary experiments and inquiries it seemed quite clear that the work should go on, the financial sanction for increased expenditure would be obtained by the appropriate Department in the usual way, and the whole responsibility would then pass to that Department. It might be that in any particular year very little would be spent, and it might be that in another year quite a large sum of money was required. As a sheer guess I should say that a total sum of £50,000 annually might be enough, and that £100,000 would certainly be enough.

2.3.7 The Royal Society Is Anxious to Help

From Sir William Bragg, FRS, letter to Admiral of the Fleet Lord Chatfield, Minister for Coordination of Defence, 13 July 1939, Public Record Office Cab. 27/712.

In July 1939, with the threat of war a constant preoccupation, the President of the Royal Society expressed the Royal Society's willingness to form a coordinating committee, to be responsible to the Committee of Imperial Defence, in the event of war breaking out.

My dear Lord Chatfield,

During the last twenty-five years there has been a notable improvement in the utilisation of scientific men in the service of the State, and in the application of the results of scientific research by various Government departments. I believe, nevertheless, that further improvement is still possible, and I would suggest for this purpose that more use could be made of the body of scientific knowledge available through the Royal Society. I do so now in view of the present urgency in national preparations for defence.

The research organisations belonging to separate departments may easily, to a greater or less extent, become isolated from one another and from the main stream of independent scientific thought and discovery. Being in each case devoted to a particular service and employing a particular group of men, they may, by reasons of secrecy or departmentalism or merely lack of contact, be unaware of advances made in other departments or by scientific people in the world at large. Now it is an outstanding fact that many of the most important additions to and applications of science are due to the combination of knowledge and experience from widely different sources: isolation and departmentalism, therefore, are strenuously to be avoided.

At the present time, it would appear that the most important problems before the Government are those which are directly or indirectly connected with defence; and that search should be made for all possible means of quickening and improving the application of science to those problems. Further provision of means of contact between the various departments though desirable in itself, would not alone ensure the best result: contact with the main body of scientific workers and of scientific research is also required and is possible only through some independent organisation. I suggest therefore that a very small group of representatives of the various branches of science might be formed from leading members of the Royal Society, who are naturally in close touch with the main body of its Fellows. The duties of these representatives would be:

(1) to make themselves aware of what the various scientific organisations under the Government were doing; whenever possible to draw attention to knowledge that might be used, or to men who had special ability to deal with any particular matter, and to report thereon to the Committee of Imperial Defence;

(2) to advise the Committee of Imperial Defence, when desired, as to the best means of obtaining scientific advice or information on any matter, and to report in general on any deficiencies of which they became aware;

(3) of their own motion to bring to the notice of the Committee of Imperial Defence any opinion, suggestion or scientific result which might seem to be important to the purposes of that Committee.

In order to fulfil these functions it would be necessary that the proposed committee should be allowed to learn what the departmental scientific organisations were doing, and to acquaint themselves to some degree with the practical aspects of the problems involved. The committee would have no executive powers and would not directly originate or supervise research in connexion with any Department. Its function would rather be to act as a channel of communication between the Government and the scientific community.

Although I am acting on my own responsibility in writing this letter, since the Council of the Royal Society will not meet again till October, I have no doubt whatever from discussions which have taken place at past meetings that I am saying what the Council would wish me to say.

Details as to the composition of the proposed committee and other incidental points could be considered later, if and when you thought it would be well to go further with the matter. I would suggest at once, however, that the committee should be limited to not more than six persons, and that these should not be directly in the employment of any department of the Government.

I know that the Royal Society would be proud to extend in this way its traditional role of cooperation with the Government.

2.4 FREEDOM AND PLANNING

2.4.1 The Planning of Research

From *Nature*, 28 July 1934, vol. 134, pp. 117–19.

In this leading article the journal put forward a reasoned case for the utilisation of scientific knowledge and expertise in the solution of economic and social problems.

Although the social reactions of science are now widely realised and the dynamic nature of science is also perceived, the idea that society itself is dynamic and not static has yet to be grasped. Once this fundamental conception has been realised by the general populace, effective attempts can be made to utilise the scientific method and outlook to release our social order from many of the disorders which it has incurred. The attention which is to be paid at the forthcoming British Association meeting in Aberdeen to the relations between the advance of science and the life of the community is definite evidence that the idea is gaining ground, and although speakers at that meeting may feel they are 'preaching to the converted', the consequent focusing of public opinion on the subject cannot be other than helpful.

Among other attempts in recent months to face these issues and to stimulate discussion may be mentioned the further statement on 'Liberty and Democratic Leadership' issued early this year over a large number of representative signatures, which referred particularly to housing, the stimulation of consumption and the organisation of distribution, and the survey of scientific research in relation to social needs described by Professor Julian Huxley in his recent book *Scientific Research and Social Needs*. Apart altogether from his valuable account of research activities in progress, Professor Huxley poses a number of fundamental questions which require attention before we can outline any adequate programme of research in relation to social problems. Something much more than scientific research in the narrow speculative sense is required: we need also the scientific spirit and method in the shape of careful planning.

The map of scientific research which Professor Huxley attempts to draw is in itself an important preliminary to such planning. It reveals at once the lopsided development of the scientific structure of Great Britain and the lamentable neglect of the sciences dealing with man. The imperative necessity of organising research less from the production side and more from the consumption end towards the needs of the individual citizen also emerges, and these two factors alone throw a flood of light on the real causes of the displacement of labour or technological unemployment.

If science is to fulfil its function in the modern State, we must, in fact, regard it as a social activity and not as something apart from the rest of human life and interest. Not only is sharp distinction between pure and applied science no longer possible, but also the scientific movement as a whole requires scientific study, and its activities must be planned as much as any other social or industrial activity, if the maximum results are to be obtained and its resources wisely exploited. This planning of science must precede the wider participation of the scientific worker in social activities. Through it must come the assembly and exploration of the scientifically ascertained facts in neglected fields, upon which alone wise action can be based.

It is probably at this point that such organisations as the Royal

Society, the British Association, the British Science Guild, the Federation of British Industries, the Association of Scientific Workers might render valuable service. Through their efforts it should be possible to map out the scientific resources of the country, and make authoritative recommendations for the re-orientation of these resources and for the attack on neglected problems of outstanding importance. An important instrument in this respect would obviously be the newly reconstituted Parliamentary Science Committee. . . .

The proper application of existing knowledge could at least double the production of food in Great Britain and raise world production to a level which would provide the population with a sufficiency of the right types of food to ensure full health and growth and energy for all.

Here, as in such questions as adequate housing, town and country planning, the utilisation of scientific results involves economics and politics. Without the large-scale planning of industry, science is liable to cause as many difficulties as it resolves. There is all the more reason therefore for applying scientific methods not only to technology and production but also to the organisation of particular industries and to the economic life of the nation as a whole. Any subject is capable of examination by the scientific method, and consumption is just as much a problem for scientific research as is production.

The idea of regarding society itself as a proper object for scientific research is new to many, but is quite definitely forced on us by such surveys as that carried out by Professor Huxley and the situation it reveals. Moreover, the scientific worker can scarcely be in any doubt that a scientific attitude to social questions is better than an unscientific one. There are many problems presented in education, the penal system, public health and industrial welfare, in which a proper supply of scientifically ascertained facts is an indispensable preliminary to wise action. Notably does the study of population with the view of controlling it offer attractive possibilities.

The merest glimpse of the possibilities of improving the quality of human life in this way which emerges from such a survey should be sufficient incentive to the mobilisation of scientific forces to this end. To fill in the gaps which exist in research by national direction and planning of research is a first step, and may demand, as suggested by Professor Huxley, the creation of a social advisory committee and research council corresponding to those responsible for planning and financing research in the economic field. Such a council would not only be able to plan out the lines of an adequate campaign of research, but also would assist in obtaining the necessary supply of research workers trained in the social sciences by modifying both the distribution of scholarships awarded in different branches of science and the science curricula in schools and universities. . . .

The outstanding progress in every field of human activity and happiness, which is really within our grasp if science were applied in the international scale as thoroughly and efficiently as it is at present within the limits of a single business or a single industry, holds out every inducement to overcome the difficulties which private profit or national sovereignty present. After all, if the form and direction of science itself are largely determined by the social and economic needs of the place and period, even in the international sphere science is influencing the world structure. Here as elsewhere it is making for the breakdown of the system which gave it birth, and demanding the creation and development of a new order in which the needs of mankind can be more effectively served. The conception of science as a social function intimately linked up with human history and human destiny, moulding and being moulded by social forces, should summon forth from scientific workers something of the energy required to translate into policy and action the knowledge acquired by their work. Such energy will find its expression alike in the discharge of their own civic responsibilities and in sharing with their fellow citizens both this vision of the new and greater social possibilities if that knowledge is sincerely and courageously applied, and the faith that human reason by using wisely the scientific method can give us the control of our destiny.

2.4.2 Pure Science and Politics

From P. M. S. Blackett, FRS, in *The Frustration of Science* (Allen & Unwin, London, 1935), pp. 130–3.

This collection of essays brought together the views of several leading scientists and commentators on various aspects of 'planned science'. In the concluding essay Blackett argued that the ideal of 'pure science' was a myth: that science was not only dependent on economic goodwill but also on the whims of government and society.

There are, of course, many scientists who believe that it is not good that scientists should concern themselves with social matters. In the interests of science they should leave morals and politics alone. I disagree. For I am convinced that the struggle between rich and poor, between property owners and the working classes, is a struggle with which they are closely concerned. Though probably few of my scientific colleagues agree with me, I believe that it is not at all irrelevant to science, but on the contrary of enormous importance, which of the two, the worker or the owner, dominates the State. It does not need much reflection to see how very closely the pursuit of science is bound up with social organisation. For one thing, a great part of the money available for research today comes from either the Government or industry. How much money there is – and modern research requires a great deal of money – depends on how much the Government and industry believe that science is likely to be advantageous to them. It is thus the social needs of the time which lead to one branch of a science being vigorously prosecuted and another branch relatively neglected.

Those who claim that science can be aloof are usually thinking of what is often called 'pure' science – but a clear distinction between pure and applied science is impossible to draw, and, though some of the more abstract branches of a science may be sometimes temporarily immune from political matters, such immunity is very superficial. No science, however abstract, has immunity today in Germany from the political environment, nor in Russia, nor anywhere else. The important question is whether the social environment is favourable or unfavourable to science. It is obvious, therefore, that science, on the scale on which it is pursued today, is an integral part of the social organisation of the country. A modern State is a highly complicated piece of machinery, and the scientist, wherever he works, is part of the machinery. If society wants technical progress, society will endow science. If it doesn't, it won't. It is, of course, the more practical branches of science which are most quickly influenced by social needs. But the pure and applied branches are so closely intertwined that it is quite impossible to discourage the first without also discouraging the second.

If this view of the intimate relation of science and social needs that I am putting forward is true, what must we anticipate for the future? What is the likely direction of social change, and how will these changes affect science and technical progress? The political climate of the nineteenth century, under which modern science grew up, was Liberal and ever-increasingly prosperous. The great advances of science in that period had a background of industrial development and technical progress. The relation between the two was very intimate. Industrial developments stimulated scientific research. The discoveries of science then gave rise to new industries, and the new industries made more scientific development possible. Now that the whole structure of liberalism and free trade is collapsing all over the world like a house of cards; now that most of the western world is no longer getting more but less prosperous, how will science be affected? To answer this, one must try and understand what gale it is that is blowing down this card-house.

There is a prevalent theory that the collapse is due to the stupidity or wickedness of politicians. This theory seems to be the modern counterpart of the historical theory of the school books that history is mainly the result of the arbitrary action of statesmen. If only politicians and statesmen were more intelligent or less wicked, then wars wouldn't happen, or, if they did, at least all the ghastly muddle of the postwar periods would be avoided. Some of those who believe this theory are apt to look to the scientist for salvation. They contrast the achievements of the scientists in their field with the failure of the politicians in theirs. And they conclude that the scientist should come out of his laboratory and turn his gifts of honest inquiry and objective judgement to help to put right the mess left by the politician. Such views are held quite widely, and not only by scientists.

Of course, it is perfectly clear that any such hope is doomed to disappointment. Scientists, if in the position of politicians, would act like politicians. They could only contribute something important to the technique of government if there is such a thing as an objective, disinterested, and so scientific, attitude to political questions. But no such attitude is possible at present, either for the scientist or for anyone else. The scientist cannot be above the battle of politics, because the conditions under which he works are directly affected by political matters. Scientists are only the instrument by which society attains a given end, that of technical and scientific advance, just as munition workers are the instrument by which society acquires its weapons of war. Scientists can influence social affairs in minor ways as other people can, but to achieve anything of importance they must throw in their lot with one or the other of the main contending forces. In fact we cannot look to the scientists for salvation.

2.4.3 Pure and Applied Science

From Julian Huxley, *Scientific Research and Social Needs* (Watts & Co., London, 1934), pp. 20–3.

In this chapter Julian Huxley (*J.H.*) and Hyman Levy (*H.L.*) discuss science in terms of the economic dependence of the scientist on government and the distinctions between university and industrial research.

H.L. It does not seem to me that science becomes 'pure' because there are individual scientific workers whose personal motive in carrying through investigations is that they desire simply to extend the boundaries of knowledge. The existence of such a motive does not necessarily enable them to lift themselves outside their historic social epoch, but it may mean that they will concentrate their attention on problems more remote from direct application. Of course, the conclusions of science must reach beyond the limits of the social system that gave them birth, for the simple reason that science concerns itself with a study of the material physical world, and that physical world exists objectively and irrespective of particular social systems. Science, however, does not cease at discovery. It is also concerned with application, and the applications are to the systems of society in being – British, German, Russian, or American. Moreover, since scientists, like other workers, have to earn their living, it seems to me that to a large extent the demands of those who provide the money will, very broadly, determine the 'spread' of scientific interest in the field of applied science. It is from there that the driving force is exerted on the scientific movement. I know of no scientist who is so free that he can study absolutely anything he likes, or who is not restricted in some way by limitations such as the cost of equipment. With most of what you have said, however, I agree, but I think it has to be seen in this setting.

J.H. Well, I think we might see how far we agree now, after all this argument. How will this do as a formal definition? Science in the modern sense is a body of knowledge which has been tested by experiment. Historically it has grown as a result of several factors, which affect man both as an individual and as a social animal. These factors are: first, our need to exercise some control over the forces of nature; secondly, our impulses of manipulation, curiosity, and our urge to understand man's place in the universe; and thirdly, the pleasure we get out of the use of our faculties in the process of observing, understanding, and changing nature. Would you agree to that?

H.L. Yes, I think that will do, although you are still thinking of science primarily as a body of knowledge. But now, agreeing that science and the scientific movement have emerged out of the growing needs of society, we ought to study it as a movement, to examine what stress has been placed on the various aspects of it – for example, the purer as opposed to the more definitely applied. What, for example, settles how much money shall be devoted to scientific work in the Universities in comparison with severely technological work outside of them? Is there any deliberate study and control of the scientific movement as a whole, or does it just develop chaotically?

J.H. Of course, that is a very difficult question – so many factors are involved. For one thing, so much of the work done at the Universities, I quite agree, interlocks with practical applications that one can hardly draw a sharp line between the two fields.

H.L. Would you, then, agree that the Universities and other purely academic institutions are doing work essential to industry which industrialists do not or will not do for themselves? For example, Faraday's electromagnetic discoveries and his investigations of the constitution of benzene, conducted at the Royal Institution, were ultimately accepted by industrialists, but the work was not initiated by them, fundamental as it was. It had to be done at an academic institution.

J.H. That, I think, is certainly true. The industrialists of Faraday's day did not even see the possibility of applying it for quite a number of years. Or we might take the researches on heredity begun by Mendel and carried on for a long time almost entirely in University laboratories, on more or less useless animals like flies and mice and shrimps. But they are now finding important and useful applications in plant and animal breeding.

In the present condition of world affairs, it looks definitely as if industry is on the whole unwilling, and apparently unable, either to provide the broad scientific background of research out of which new applications grow or to undertake large-scale and fundamental investigations which do not promise fairly immediate returns. On the whole, it is fair to say that the Universities provide the background, and Government institutions (like the National Physical Laboratory and other branches of the Department of Scientific and Industrial Research) carry out the long-term investigations.

H.L. So you agree that Universities, whatever else they are doing, are unconsciously playing their part in assisting industrialists to carry on their business?

J.H. Yes, that is so. Of course, the Universities, like any other social institutions, cannot help doing something to serve the ends of the society in which they have grown up. But helping industrialists is only one side even of this aspect of Universities. They may help to cheapen production so that prices can be brought down and also help to stimulate new inventions, and so to cater for needs that have hitherto not been satisfied.

H.L. Yes, science has been used in this way, but even this analysis of yours is surely incomplete. There is a real distinction between two possible ways in which science operates. First, science may serve certain social and individual needs directly, by stimulating our intellectual and philosophical interest. It may expose the false basis to many of the beliefs we have inherited from the past, and provide us with assured knowledge on which to reconstruct our view of life and of society. It assists, in fact, to sharpen our critical sense and to enlarge our outlook. Secondly, however, science comes to society indirectly. It may be used by those who have made it their business to cater for more immediate practical social needs. Before the results of science get to society by this route, it has to be worth these people's while to use it. For the moment, however, we will leave that.

2.4.4 Science, Economics and Society

From J. D. Bernal, *The Social Function of Science* (Routledge, London, 1939), pp. 320–4.

Bernal's book presents a Marxist argument for the relationship between science, economics and society. It was immediately recognised as a work of major importance in the debate on freedom and planning in science. In this extract Bernal argues that once science is seen as a vital force in human advance it will be supported by the state, but while it is starved of support it will never play its proper role in the development of society [cf. 4.2.1].

Economic Nationalism and Planned Science. – Of increasing importance in the finance of science in modern States is its national economic aspect. Indeed, it is almost to this alone that we owe the preservation, in certain States, of any science at all. In Germany, for instance, the whole atmosphere of public life is antagonistic to the spirit of science. Blood and soil are considered more important than intelligence, and yet it is reluctantly admitted in a modern world that blood and soil are not enough to secure national honour and national freedom. Science is required for two things: the perfection of the war machine, and, what is merely another aspect of the same thing, the perfection of the national economy in the direction of making it self-supporting. Yet, though this is the most extreme example, the same tendency is plainly present in all other capitalist countries. It has stimulated in Britain, for example, the formation and continuance of the Department of Scientific and Industrial Research. The effect of national economic pressure on science is largely to drive applied research in two directions: one, into the heavy industries mainly concerned with armaments, particularly the metal and the chemical industries; and the second, to a lesser degree, into the problems of food production and preservation, thus exaggerating the disproportion that already exists between the physical and the biological sciences. This would, of course, not be so marked if the food research were of a more biological character, but here we come across one of the contradictions inherent in modern politics, namely, the political concern for primitive methods of agriculture which always runs parallel to economic nationalism. It is necessary to keep agriculture primitive because the conservatism of landlord and peasant must not be disturbed, as it is on them that depend the strength of civil reaction and military manpower. As a result, enormous chemical ingenuity will be expended on producing synthetic foods to save the relatively small administrative and political changes necessary to introduce a rational agriculture. The researches on food preservation turn out in practice to result in advantages far more to the middle men, organised in large distributing trusts, than either to the direct producer or to the consumer. Yet in an indirect way the development of science in the interests of economic nationalism can be advantageous because it indicates, for the first time, some conception of an organised scientific attack on problems of interest to the community, and suggests that in better times such organised research might be turned from its present purposes of preparing for war to those of benefiting the community.

The Freedom of Science

This sketch of the possibilities of science in two different types of social and economic environment may serve to show what are the requirements of a social organisation in which science can play its full part. It is essentially a question of a wider aspect of the freedom of science. The freedom of science is not merely the absence of prohibitions or restrictions on this or that research or theory, though in certain countries today science has not this elementary freedom. The full freedom of science goes much further. It is useless to permit a research if at the same time the funds to carry on that research are unprocurable. Lack of means fetters science as effectively as police supervision. . . .

Frustration. – The full development of science is only possible if it can play a positive and not merely a contemplative part in social life. This is certainly what science was doing in the great days of its advance, in the seventeenth and in the early nineteenth centuries, when capitalism provided for the first effective utilisation of natural forces. But more and more today the utilisation of science is being cramped and devoted to base ends. The lack of freedom and distortion of science in turn reacts on its internal development. Where a great tradition has been built up it is still possible for science to follow out the lines of that tradition, but elsewhere, as in the biological and sociological sciences, advance has been definitely held up. Science divorced from the effective life of its time is bound to degenerate into pedantry.

The general problem of the finance of science is thus seen to be far more social-economic than purely scientific in character. Once science plays a recognised part in social advance the problem of raising money for it adequately under a rational plan should not be difficult. The total amounts required are so small that, except in times of acutest crisis or in reconstruction after destructive wars, there should be no difficulty in finding ample and more than ample funds for scientific research. Once science is so organised that its benefits can flow rapidly and directly to the public at large, its value will be so obvious that there will be no difficulty in setting aside the 1 or 2 per cent of the national income which will represent as much as it is likely to absorb for the next two decades and from five to twenty times the amount it absorbs today in most capitalistic countries. There is so much to be done that the limiting factor will not be the amount of money available but the number of men who can use that money. Science has the prospect of being fully employed until human needs are met in a way of which we at present can form no picture.

Science needs Organisation. – We have now considered in its different aspects the general problems of science organisation. The discussion has necessarily been somewhat academic because we have been concerned with forms which, because they exist only in a possible future, do not lend themselves to concrete examples. In such a treatment only measurable factors can be taken into account, but it is the unmeasurable factors that are far the most important. No organisation, however well thought out and however integrated with the general social scheme, can be of any use if it does not represent the effective desires of the people who are to work the organisation. It is largely then from the attitude of the scientists themselves and of the public towards science that we can gauge the possibility of the success of any reorganisation of science. It would be idle to deny that there has existed in science up to the present a distrust of any organisation whatever. But this distrust is founded partly on the old tradition of the freedom of science from the obscurantist restrictions imposed by the Church and by scholastic universities and partly on the more immediate experience of state-regulated science. As to the first, this dwelling on past struggles of science has too often served to obscure the real danger of the present: no longer the suppression of science as a whole but its exploitation in detail. The freedom of science needs to be considered in its modern aspect as freedom to act and not merely to think. For this organisation is necessary, but organisation of science does not and cannot mean, if it is to be effective, the type of organisation that has been taken over in unthinking fashion from business or civil administration. To submit science to such discipline and routine is certainly to kill it. The great section of science that suffers from it at the moment is in fact effectively dead. But organisation need not mean such discipline and routine. It can be, as we have tried to show, free and flexible at the same time as remaining ordered. If it retains as its central core democratic spirit expressed in democratic forms, no organisation of science will be able to lose touch with the corporate feeling and the desire for knowledge and human betterment that is inherent in the effective progress of science. If we are to have an organisation of science it must be largely built up by the efforts of the scientists themselves. . . .

Scientists and the People. – The setting up of a scientific organisation, however, cannot rest with scientists alone. The scientists cannot force their services on society; they must form part of a willing and conscious partnership between science and society. But this implies far more than an adequate appreciation on the part of the non-scientific public of the achievements and possibilities of science. For its full effectiveness it requires also a society economically organised so that general human welfare, and not private profit and national aggrandisement, is the basis of economic action. With such economy the scientists, possibly more than any other of the relatively wealthy sections of present society, will find themselves in agreement. For science has been at all times a commune of workers, helping one another, sharing their knowledge, not seeking corporately or individually more money or power than is needed for the pursuit of their work. They have been at all times rational and international in outlook and thus fundamentally in harmony with the movements that seek to extend that community of effort and enjoyment to social and economic as well as intellectual fields. Why this fundamental identity is not as yet fully realised by the scientists or by society will be discussed in a later chapter.

2.4.5 Counterblast to Bernalism

From John R. Baker, *New Statesman & Nation*, 29 July 1939, vol. 18 (NS), pp. 174-5.

In this article Baker, the Oxford zoologist, rejects the arguments advanced by J. D. Bernal in *The Social Function of Science*.

'. . . *a higher degree of Reputation is due to* Discoverers, *than to the* Teachers *of* Speculative Doctrines . . .' (Sprat, 1667.)

Bernalism is the doctrine of those who profess that the only proper objects of scientific research are to feed people and protect them from the elements, that research workers should be organised in gangs and told what to discover, and that the pursuit of knowledge for its own sake has the same value as the solution of crossword puzzles. Professor Bernal will no doubt permit the immortalisation of his name by the introduction of this new word into the English language.

Opponents of Bernalism, well aware of the way in which great discoveries have been made in the past, are appalled at the thought that if once Bernalists get power the fount of discovery will be dried up. If science is left free, incalculable benefits for human welfare, material and mental, will accrue as they have accrued in the past; but it would be as undesirable to try to force organisation on scientists as on composers, artists or choreographers. It is true that music can be composed under duress. One knows that Mozart, whose antipathy to the harp was only equalled by his positive dislike for the flute, was persuaded to write a very tolerable concerto for these two instruments. Yet even de Guisnes would not have been able to get him to submit to the organisation which Bernalists seek to impose, far less to confine himself to the composition of music of the greatest economic value (i.e. presumably, that which would give employment to the greatest number of musicians and musical instrument makers).

Let the reader try to imagine Charles Darwin told off by the Bernalist-in-charge to organise a gang of Ph.D. students. The mind reels before the thought. Among other famous workers in the biological sciences who showed by their lives a particularly obvious antipathy to the principles of Bernalism one may mention T. H. Huxley, Russel Wallace (co-founder with Darwin of the theory of evolution by natural selection), Mendel (founder of the Science of Heredity), Fabre (student of insect behaviour), Jenner (student of the cuckoo and discoverer of vaccination), Schwann (founder of the cell-theory), Linnaeus (inventor of the scientific method of naming plants and animals), van Leeuwenhoek (discoverer of blood-capillaries, spermatozoa, etc.), and Malpighi (discoverer of the air-cells in the lungs, etc. etc.). These are men whose names will be revered when even the derivation of the word Bernalism has become obscure.

Many famous workers in other sciences besides biology could never have been forced into the Bernalist rut. Supreme as an example of these was Cavendish, discoverer of the fact that water is formed by the combination of the two gases which were later named hydrogen and oxygen. Could Cavendish have worked in a gang under orders and confined himself to economic research? So far was he from being capable of collaboration, that he could not bear the proximity of others and had his library built four miles from his house, so that those whom he permitted to borrow books should not come near him when they did so. That he worked mainly for his own enlightenment is shown by his never bothering to publish his monumental researches in electricity, which were only brought to light after his death. Newton, it will be recollected, was almost equally unconcerned as to whether others got to know of certain of his most important discoveries. This was indeed regrettable, but it shows that geniuses have pursued knowledge for its own sake; and in doing so they have thrown off much that has been of tremendous significance for the practical affairs of mankind.

Science has two functions – to serve human welfare in material ways, and to increase man's capacity to comprehend the universe. To many who are contemplative there is nothing more worthwhile in life than the increase of knowledge for its own sake. To pretend that this is 'escapism' and comparable to interest in cross-word puzzles is nonsense. No amount of solving puzzles increases anything worth increasing; but every discovery in science separates falsehood from truth and makes an accretion to that vast body of demonstrable knowledge whose possession is the most valid criterion of distinction between cultured and savage communities. There are those who are unrewarded by the contemplation of nature, just as there are those who find nothing in music or art. Nevertheless, knowledge and music and art are among the ultimate things in life for many people who regard just keeping alive and healthy as merely means to an end.

One of the keenest and most satisfying pleasures to many people is the possession of some understanding of the landscape which lies before them when they travel, whether in this country or abroad. This may be taken as an example of how science can serve as an end in itself. To walk in an unfamiliar part of England with Trueman's *Scenery of England and Wales* and Tansley's *Types of British Vegetation* in one's haversack is to experience something which is denied to those whose appreciation of landscape lacks understanding and is entirely aesthetic. When the walker opens his haversack to eat, he has the Bernalists' approval, because eating is useful. When he opens it to fill his mind and not his stomach, does he necessarily sink to the level of the solver of the crossword puzzle? Is he really to fall down and worship the grocer who sold him his lunch, and pretend to ignore his debt to Professors Trueman and Tansley? Whether subsequently he does or does not make 'use' of the added knowledge which his study of scenery has given him, the study is worth while as an end in itself. In the American Ambassador's memorable words, spoken at Manchester a few days ago, 'that desire to find out, which is unencumbered by any desire for a practical cash return, has given us many of civilisation's great blessings'.

Perversely enough, the keenest Bernalists are often those whose own research has been the most meagre in supplying the material wants of man. Let them devote themselves for a few years to the scientific investigation of some really practical matter, such as making two blades of grass grow where one grew before, or one baby grow where two grew before. They might then be more useful members of society and waste less of their time in yapping at those who are of a more contemplative nature than themselves. What a scientist ought to do is an ethical concern for the judgement of his own conscience. Those to whom one listens with respect when they speak of verifiable matters (e.g. in crystallography) compel attention much less inevitably when they try to lay down the law on moral issues.

There is room for all in science. Let the gangsters work always in gangs and order one another about and improve whatever it may be that they tell one another to improve. Everyone will be thankful for their services. Let there be freedom, nevertheless, for those who lack the gang instinct but possess that insatiable curiosity, sneered at by Bernalists as puerile, which is the source of all real advance in science, whether pure or applied. If these can be left free and unhampered, science and its applications will flourish. Let Bernalists remember the words with which Faraday answered one of their number who inquired the use of his experiments: 'Madam, will you tell me the use of a new-born child?' Let them ponder also the magnificent lines which Prebendary Sprat wrote of their forerunners in 1667: '. . . they [the Bernalists] are to know, that in so large, and so various an *Art* as this of *Experiments*, there are many degrees of usefulness: some may serve for real, and plain *benefit*, without much *delight*: some for *teaching* without apparent *profit*: some for *light* now, and for *use* hereafter; some only for *ornament*, and *curiosity*. If they will persist in condemning all *Experiments*, except those which bring them immediate *gain*, and a present *harvest*: they may as well cavil at the Providence of God, that he has not made all the seasons of the year to be times of *mowing, reaping*, and *vintage*.'

3 Central Organisation: The Second World War and After

In this chapter we continue the story of the central organisation of science and technology from the second world war up to, in effect, the end of 1966 when the establishment of a new Central Advisory Council for Science and Technology in the Cabinet Office was announced. This marked the culmination in the postwar planning machinery for science and technology.

The innovations introduced by the government under the pressures of war did not go unregarded when peace came. The Prime Minister had been advised throughout by Lord Cherwell and the Cabinet had been supported by a Scientific Advisory Committee.[1] In 1946 the new Labour government was advised by Sir Alan Barlow to build on the wartime experience and set up an Advisory Council on Scientific Policy (ACSP). This they did [3.2], together with a new Defence Research Policy Committee; Sir Henry Tizard was chairman of both.

The decision to set up the ACSP fell far short of the much more radical suggestions of bodies such as the Association of Scientific Workers (AScW). The AScW rejected the idea of a Science Ministry but advocated a Central Scientific Office at Cabinet level [3.1].

In retrospect the ACSP can be seen as only a pale reflection of this; its function was advisory only, not executive and its secretariat probably never measured up to the scope of the Council's field of interest [3.16].

Despite the rapidly increasing public expenditure on science after the end of the war (civil science accounted for £5·2 million in 1945–6 and £34·3 million fifteen years later) and such achievements as the beginnings of nuclear power generation and space exploration, it was not until 1959 that science first emerged as a

[1] Scientific Research and Development, Cmd. 6514, April 1944.

central political issue [3.3]. The victorious Conservatives in the general election of that year then redeemed their election pledge by appointing a Science Minister. (Labour, on the other hand, promised 'a scientific and technical planning board' to assist 'a senior Minister of high standing' with general responsibility for science [3.4].) Their appointee, Viscount Hailsham, had already been Lord President of the Council since 1957 with the traditional responsibility for the research councils. In 1959 he acquired the broader remit of Minister for Science and eventually, in 1963, the responsibility for the universities too.

The new arrangements were a source of controversy, in part because it was held that the new machinery for science was inadequate for the tasks facing it, and in part because of the Minister's expressed views [3.5]. Criticism came not only from the obvious place, the Parliamentary Opposition [3.6], but also from Conservative backbenchers and the Federation of British Industries, reflecting the reformist spirit abroad [3.7–3.9].

The criticism took many forms: that the Cabinet's advisory structure should be strengthened, that the resources of the Ministry of Aviation should be switched to more general industrial application, and that new ministerial arrangements be made. The Robbins Committee on higher education offered its own suggestions based on different premises [3.11]. But the key event was the reporting of the Committee of Enquiry into the Organisation of Civil Science (the Trend Committee) at the end of 1963 [3.12]. This investigation stemmed in large measure from the ACSP's own admission that its task had grown too big for it to cope with, that it lacked 'the machinery for arriving at decisions on major priorities' (Annual Report 1960–61, Cmnd. 1592, January 1962, para. 39).

Much comment, most of it adverse, was directed at this report [3.13, 3.14] which advocated, *inter alia*, the dismantling of the Department of Scientific and Industrial Research and the creation of two new research councils and an autonomous Industrial Research and Development Authority [4.3.5, 4.4.4]. The Committee also advocated a substantial strengthening of the Minister for Science's role, the severance of the Privy Council link with the research councils, the enlargement of the Minister's Office and the setting up of a new advisory body to assist the Minister: all implicit statements that the existing machinery was defective.

In February 1964 a 'federal' Department of Education and Science was created and the government also announced that it would implement most of the Trend Report recommendations. A general election followed in October with a change of government.

The Labour Party made science and technology into a major

issue in the run up to the 1964 general election, building upon such issues as their accusations that the Conservatives had misdirected the country's scientific potential, failed to plan for the future and were allowing the country's already inadequate manpower stock to drain away [3.10]. However, some commentators could see little merit in either party's proposals [3.15]. With their victory the Labour Party were enabled to introduce their promised Ministry of Technology 'to guide and stimulate a major national effort to bring advanced technology and new processes into industry'.[1] Under the same piece of legislation[2] the two research councils for science and the environment proposed by Trend were created. The new government retained the Conservatives' Department of Education and Science but abolished the old Privy Council link[3] and made the research councils responsible to the Secretary of State. He was now advised by a Council for Scientific Policy (in effect replacing the former ACSP), whose remit ran to civil science alone [3.17]. For technology there was an Advisory Council on Technology to aid the Minister. The division looked neat on paper but was a cause of worry to some [3.18, 3.20]. The advisory structure was bridged by a Committee on Manpower Resources for Science and Technology (replacing the former Committee on Scientific Manpower of the ACSP). Finally in October 1966 the Central Advisory Council for Science and Technology was formed as advisers to the Cabinet itself [3.19].

3.1 Central Organisation and Coordination of Scientific Research

From Association of Scientific Workers, *Science and the Nation* (Penguin Books, Harmondsworth, 1947), pp. 174–8.

The Association, a scientists' and technicians' union and also a scientists' pressure group, numbered some key figures in the development of British science policy among its membership, e.g. J. D. Bernal and P. M. S. Blackett. This is an extract from one of its major policy statements.

The suggestions of this book for the postwar organisation of science in this country are based on the need for a body capable of forming a general broad policy for directing scientific research

[1] Labour Party Manifesto 1964.
[2] Science and Technology Act 1965.
[3] See the Introduction.

to the problems most important for the country, taking instructions from and reporting to the Cabinet. The need for central coordination of our scientific effort is now generally recognised as essential to efficient reconstruction of our national economy. The experience and successes of the war, when the National Government found it necessary to coordinate scientific resources, and the election in July 1945 of a Labour government pledged to reject *laissez-faire* as a national policy, have prepared the way for such a development.

The urgency of adequate coordination is particularly emphasised at the present time by:

(*a*) The acute shortage of scientific man-power.
(*b*) The need to establish priorities in the allocation of finance and personnel.
(*c*) The need to ensure the adequate expansion of scientific work in fields at present neglected.

The form of the future organisation must maintain and in fact increase the opportunity of the individual scientific worker, or the team of scientists, to pursue investigations freely, and should give maximum autonomy to all research units to plan their own work within broad terms of reference. At all levels there should be opportunity for scientific workers to take part in the democratic planning of their own work. The future form should combine the minimum of dislocation with the maximum utilisation of the existing structure.

The present Government is committed to a positive policy of national economic development. This requires an equally positive and parallel policy for science. Responsibility for the various sectors of the national economy rests with Ministers and hence, in keeping with present practice, each Minister requires an appropriate scientific organisation within his own department.

These needs can be achieved by an adaptation of the existing system and the addition of a Central Scientific Office at Cabinet level, with further coordinating machinery between Government, university and industrial science. An alternative suggestion for a Ministry of Science, which has been made in certain quarters, should be rejected. If all scientific work were done in a single Ministry, the rest of the Ministries would be divorced from science and the valuable responsibility of Ministers for an essential element of their own work would be lost.

Each Ministry should have its own scientific service suitably strengthened to make it fully effective. This would involve some of the Research Establishments at present responsible to the DSIR, or at least part of their work, being transferred to the appropriate Ministries, e.g. the Board of Trade, Ministries of Fuel and Power, Works, Supply, Transport, Health, etc. The effect of all this on the Government role in science would be:

(*i*) in the nationalised sections the Government would have direct control.

(*ii*) in the non-nationalised section the Government would have a policy and an interest which would be promoted through advice given directly to industry or through the Research Associations.

It would be possible for the Ministries in this way to carry out positive development work for the benefit of both the nationalised and non-nationalised sectors. Careful thought would be needed to determine how much research to leave under central government direction and how much should be departmental. The general principle might be to leave background research, i.e. research with common interest to several industries, such as that carried out by the National Physical Laboratory or general applied biological work, under central control, but to leave applications and developments with a definite relation to a particular sphere of national economy under the general direction of the individual Ministries.

A Central Scientific Officer should be established at the Cabinet Secretariat level as part of the office of the Lord President of the Council. This office should coordinate national scientific policy with the economic policy of the Government, and should be linked with other planning agencies in the Cabinet Offices and with the Central Statistical Office. Both administrative and scientific members of the staff would be needed. Some of the scientific members of this Office should be seconded from Ministries, including Supply; others should be specially appointed, e.g. seconded from the universities for a period of years.

The Central Scientific Office, together with other planning agencies in the Cabinet Offices and the Central Statistical Office, would have the function of informing the Cabinet of the progress of scientific work and of the availability of scientific personnel, of suggesting adjustments and new programmes of work, and of arranging for the necessary implementation of plans and the provision of adequate personnel.

Such a Central Office would not be suitable for dealing directly with the great variety of scientific problems that would have to be covered. Some division into subjects would be necessary. There are a number of sciences common to two or more Ministries; thus genetics serves agriculture and medicine, and the social sciences should serve practically all departments. Background research which affects more than one Ministry should be organised under Research Councils. It is suggested that three Research Councils, of the physical sciences, the biological sciences and the social sciences, would be appropriate to cover the field. These councils could grow naturally out of the bodies already existing; thus the present Department of Scientific and Industrial Research (DSIR) would father the Physical Research Council, and the present Agricultural Research Council (ARC) and the present Medical Research Council (MRC) jointly the Biological Research Council. An entirely new body would have to be formed for the Social Research Council, though this might contain elements from the economics and statistical branches of the Government. Each of these three Councils would need a central administrative and scientific staff and would have responsibility for the various research establishments working under them. They would be Executive Councils receiving an annual grant-in-aid and would have the same constitutional position as the present Medical Research Council.

An Interdepartmental Coordinating Committee will be necessary to coordinate Government science. It should include senior scientists from each Ministry and the secretaries of the three Scientific Councils, with the Chairman of the Central Scientific Office as Chairman.

3.2 Terms of Reference of the Advisory Council on Scientific Policy

From the First Annual Report of the Advisory Council on Scientific Policy 1947–8, Cmd. 7465, July 1948.

The establishment of the ACSP marked the first postwar attempt to plan science coherently and comprehensively. Its chairman until

1952 was Sir Henry Tizard, a former Secretary of the Department of Scientific and Industrial Research and a key figure in the prewar controversy with Lord Cherwell and the government.

3. In a minute which you addressed to our Chairman shortly before our first meeting, you made it clear that you expected us to initiate discussion of problems without waiting for them to be referred to us, whenever we felt we might suggest ways in which the application of scientific knowledge and experience could assist in the solution of the Government's problems. You also asked us to make recommendations on a number of specific issues of policy. These were:

The appropriate organisation for scientific research within the Government with special reference to research on building and on fuel and power;

The arrangements for securing an adequate flow of scientific manpower to meet the needs both of Government and of industry;

The appropriate form of research effort to assist the maximum increase in national productivity during the coming decade; and, as a special problem,

The appropriate organisation for oceanographical research.

7. You had asked us to consider, in particular, what should be the respective functions of Government Departments, of the Research Councils and of outside bodies (including the Boards of socialised industries) in carrying out research. We therefore devoted considerable time at the outset to discussion of the general principles which should apply to the organisation of government research and its application.

8. It soon became clear that there existed two divergent views amongst those interested in the future organisation of science within the Government. On the one hand there was the view that it was important to preserve the independence of the Research Councils in the interests both of the work and of the acceptance of the results as being clearly impartial, by industry and others concerned. On the other hand there was the view that Executive Departments concerned – the Government users of the results – should be assigned and be expected to assume a more positive role in the organisation and direction of research required for their own

purposes. We were, however, in agreement on the following general principles:

(i) The executive department should be responsible for identifying problems requiring research, settling their order of priority, deciding where the various investigations should be carried out and applying their results.

(ii) The Research Councils and particularly the various branches of the Department of Scientific and Industrial Research should, as in the past, be free to initiate background research where they thought fit, free from administrative control of the executive departments and consequently from considerations of day to day expediency. They should also undertake research at the request of the executive departments.

9. We also agreed that no attempt should be made to concentrate responsibility for carrying out all Government scientific research on the Lord President or any single Minister since this would be a step in the direction of creating a Ministry of Science, and would thus intensify, rather than diminish, the old cleavage between science, administration and policy. We expressed the opinion that the Lord President ought to assume a particular concern for such branches of scientific research as have a wider application than to the problems of a particular Department.

10. We came to the conclusion that if the executive departments were to discharge their considerable scientific and technical responsibilities as defined in 8(i) above, they would have to employ scientific staffs, corresponding approximately to the operational research units attached to the Service Ministries and Commands during the war. We, therefore, recommended that Departments should, as a rule, appoint a Chief Scientific Officer, whose functions should be to advise generally on the scientific aspects of departmental policy, and, in particular, to define the problems calling for research, to make the necessary arrangements for its conduct and to watch the application of its results. We further recommended that where such an officer was appointed, he should be provided with an adequate scientific staff capable of undertaking operational research covering the economic, technical and other factors affected by scientific advances. We also advised that in some Departments it might be advantageous to appoint an Advisory Council which

would assist the Minister in formulating the scientific policy of his Department.

3.3 A Science Minister

From *The Observer*, 20 September 1959, p. 16.

After the election, a Minister of Science will be appointed, whichever party wins – this is now clear. Since much of the nation's scientific research is supported by Government funds, the proposal that there should be a responsible Minister, instead of the present variety of *ad hoc* arrangements, seems at first very sensible.

Nevertheless, some very careful thinking will be needed before the powers and functions of the new Minister are defined. For one thing, he will inevitably find himself entangled with other interests: the major proportion of Government research expenditure, for instance, is on defence. Already involved in decisions about this are the Minister of Defence, the Minister of Supply, the Admiralty and Service Ministries, the Atomic Energy Authority, and any number of advisory committees. If the Minister of Science is made responsible for increasing the supply of scientists, he will have to work with the Ministry of Education, the University Grants Committee and others. If he tries to promote more research in industry, he will have to clarify his relations with the Department of Scientific and Industrial Research – perhaps by taking it over, as Labour intends.

There is also a more serious problem. A Minister of Science will be under great temptation to support projects which will bring a *political* return within his term of office. He will want to have some spectacular and easily explained results to show, such as a supersonic airliner or a fleet of hovercraft. This must be avoided at all costs. The first essential, in this country, is for science to penetrate more widely and thoroughly into the more backward industries. (Labour's plan for giving them more development contracts would help.) This penetration will be a slow, unspectacular process – but it will yield better value for money in the long run than anything else.

3.4 A New Deal for Science

From Labour Party Policy Statement, October 1959.

Labour policy, announced shortly before the 1959 general election and in response to the Conservatives' promise in their manifesto ('The Next Five Years') that:

> One Cabinet Minister will be given the task of promoting scientific and technological development. While it would be wrong to concentrate all government scientific work into a single ministry, this Minister for Science will have responsibility for the Department of Scientific and Industrial Research, the Medical and Agricultural Research Councils and the Nature Conservancy, the atomic energy programme, and the United Kingdom contribution to space research.

To use modern science to the full requires a carefully planned programme which must be energetically carried through by the Government.

For this reason there will be appointed in the next Labour Government a senior Minister of high standing who will be given general responsibility for scientific affairs. It will be his job to ensure that the necessary plan is perfected and carried out by and through the various Ministries concerned. He will be given the authority to do this. What matters is not whether he is called 'Minister of Science' but that he should do his job properly. The plan must include the following:

1 A further expansion of scientific and technological education, in part by founding new Universities and in part by expanding and improving the Colleges of Technology. To encourage fundamental research we shall increase the grants to Universities and consider setting up a new Research Council for this purpose.

2 More scientific training in the schools, so as both to provide recruits for the Universities and also to improve the general standard of technical knowledge in the community. This, in turn, calls for more science teachers. In time the expansion of the Universities will provide them. Meanwhile, special measures will have to be taken. We shall, therefore, institute an extensive campaign of recruitment for science teachers, including the encouragement of those who have retired (especially women who have done

168

so on marriage) to return to work. We shall also ask scientists from industry or the Universities to help in this drive by doing some part-time teaching in the schools.

3 To secure the more rapid application of the latest scientific knowledge to industry, we shall substantially increase the number of research and development civilian contracts given by the Government. We shall also make grants to individual firms for approved long-term research projects. Such assistance will be dependent on the firms taking whatever steps are necessary to improve their scientific and technical direction.

It may also be dependent on the amalgamation of a number of small firms into a single enterprise capable of supporting a research programme.

4 In order to supervise the application of science in industry we shall set up under the Minister concerned a scientific and technical planning board whose task it will be to advise the Government on the direction of industrial research and development, on the awards of research contracts and on the grants to individual firms.

3.5 Science and Government

From Viscount Hailsham, 8th Fawley Foundation Lecture, delivered on 9 November 1961.

Lord Hailsham held the posts of Lord President of the Council and first Minister for Science. His conception of his role at the head of the government's civil science machinery was a cause of controversy; in this Lecture he outlined his personal view of his office.

I sometimes read, in published prints, attractive prospectuses of what is called 'scientific planning' – with sometimes a 'science budget' or a 'scientific general staff' added, or even a demand for what is called a 'real Minister of Science' 'with real powers'.

Just how much there is in these ideas I may have to examine later on. I am now concerned to analyse the problem. The first thing is to insist that, from the point of view of planning, science cannot be isolated from politics or economics. There may or may not be a case for a national economic plan; indeed, much current political argument centres round the senses in which such a plan –

or plans – can be made for the nation or for individual industries. In such a case the scientist and the technologist would, or at least should, have to play an important part in the planning. But what is plain is that, except in the most abstract realms of pure science, there cannot be a scientific plan which differs from the economic plan or can even be isolated from it. There is no set of circumstances in which the oceanographer, the nuclear engineer and the lepidopterist meet together to discuss a common national plan which has meaning except in the context of politics or economics.

Not only can science in general not be divorced from economics or politics, but the individual sciences *can*, to some extent, be divorced from one another. True, they may each cross-fertilise the others; and equally they all form part of a vast single corpus of knowledge. So they do, but this corpus of knowledge is not purely scientific, and the relationship between the activity of one branch of science – let us say the research of a man investigating the cancer cells of mice – and that of another – say, looking at the transmission of a motor car – is not necessarily any closer than the relationship between either of these activities and the great complex of medical, industrial and engineering problems in which in each case the research is directly related. That, incidentally, is why in this country we do not have a single 'scientific general staff' so much as four closely related, but independent research councils, each of which, if one cares to use the cliché, is a general staff in its particular field.

In the ultimate analysis, the problem of control can be stated quite simply. So much depends upon science nowadays that Government cannot disinterest itself in scientific policy. At the same time, no Government, slave or free, has yet devised a means of securing that science can be done by anybody else than the scientist – and by these I mean men and women actually carrying out scientific work with laboratories and often with students, and not simply people with good science degrees engaged not in science, but administration.

A problem closely related to the questions I have been discussing is related to balance, and priorities. Here again, money is not the useful yardstick which it is in so many matters of practical administration.

In the first place, sciences cost different kinds of money. A good particle accelerator may cost several millions, and be, when completed, about the size of the Round Pond in Kensington Gardens. I am thinking of that at Cern, and of Nimrod at Harwell. I doubt whether the entire National Medical Research Institute cost as much in much less than a decade. Each would be two years' purchase of the entire MRC revenue. Other expensive items are, clearly, radio-astronomy and atomic energy, and, of course, space research. Some of the projects in the earth-sciences can also be listed. But, in the nature of things, biology employs less hardware – whether in its fundamental aspect, or in its medical or agricultural applications. That does not mean that it is less important, or even necessarily that it is being less actively pursued. But it does mean that 'big science', as the Americans are beginning to call the items I have mentioned, must, if it is pursued at all, proceed on a different scale of expenditure.

Moreover, as in other things, there are fashions among sciences, and among the scientists. At one particular moment in the history of science one particular branch appears to attract the brains and stimulate the imaginations. This, in the end, is what will attract the money, and if the money is lavished on anything else – even something intrinsically more important – it will quite probably be wasted.

The other difficulty in assessing questions of balance arises from the fact that scientists are intensely specialised in their activities and tend to have their own enthusiasms. A committee of enthusiasts for space research can be counted on to produce a report explaining the public advantage that will flow from rockets and satellites. But how far would their arguments be found convincing by a committee interested in the promotion of a supersonic airliner or a nuclear ship? And how would the biologists or the Nature Conservators react to these? Personally, I have always fought shy of anything approaching a 'science budget' – because it seems to me in the end meaningless in that context to hold up a project for cancer research because the money is needed for a new synchrotron. Both are indeed capital investments. But each is related to other things outside science more closely than either is related to the other. Nevertheless, some kind of priority is necessary – at least for items of hardware in certain fields costing more than a few hundred thousand pounds. How can we select between projects without either ignoring or accepting without question the demands of the enthusiast in any discipline? It is clear that a mechanism of some sort for measuring scientific priorities within different disciplines is unavoidable.

The fourth problem I notice is that of application.

Whatever form of Government organisation is chosen for the encouragement of scientific research, the application of scientific knowledge to industrial processes and concrete policies is something which will never be confined to research institutes or universities.

The first thing which I have really got to try and get across – and which so far I have failed to get across – is that the tidy, ambitious and grandiose schemes which I am constantly being invited to adopt for the aggrandisement of my office, and the enhancement of my personal reputation, would in effect be reactionary and restrictive of my true activities. In Government it is important not so much that there should be a Science Minister as that all Ministries should learn to regard the application of science, each in its own particular sphere, as one of its main responsibilities. Education, Power, Transport, the Post Office, the Defence Ministries, Housing, Agriculture – all such Departments must have scientific staff of some kind as part of their organisation if they are to function at full effectiveness. Research should be carried on, and applied, by the nationalised industries and private firms. There is not, and there cannot be, a central organisation to carry all this out; the function of any central organisation must be to stimulate and encourage this work in others, to fill in gaps, to conduct generalised researches not apt to the functions of narrower institutions, to create organisations and institutions where these are lacking. The function of Government in this sphere is patron, not employer; in a sense, indeed, it is that of an impresario.

3.6 The Minister for Science

From the Labour Party, 'Science and the Future of Britain', March 1961, pp. 40–2.

The Labour Party were highly critical of the administrative changes introduced by the Conservatives after the general election of 1959 and in particular of the role of Lord Hailsham.

To be effective, the Minister for Science should be a senior Cabinet Minister, of sufficient standing to carry great weight in political and economic discussions. He should have responsibility for the Department of Scientific and Industrial Research, the Medical and Agricultural Research Councils, the Atomic Energy Authority, the Nature Conservancy and the National Research Development Corporation (now attached to the Board of Trade) as well as a Central Scientific and Technical Register of technical manpower. His should be a full-time job and not, as has usually been the case with the Lords President of the Council, something coming second or third after other duties such as Leader of the House or Party Chairman. It is clearly preferable that he sit in the Commons, but if he is a Peer he should have a single representative in the Commons, with the standing of a Minister of State.

His secretariat need not be large in numbers but must be of a very high level of ability. It should be headed by a forceful Permanent Secretary able to hammer home the Department's case at administrative level; he should be supported by a leading scientist-administrator as Director-General.

The Minister must work out a scheme for the coordination of science, in fullest consultation with those concerned in science. The development of science may require new forms of organisation – the Atomic Energy Authority is an example of this. Research workers, engineers and technologists, teachers (and not only science teachers) industrialists, civil servants, political scientists and politicians have important points of view on this problem. Moreover, any coordinating body, once it is in operation, must continue to consult the professional organisations and trade unions of the scientists themselves.

There are two major omissions in the present arrangements for the Ministry of Science. The truth is that the new Ministry has no effective power, except over the DSIR, in the development of scientific policy. Major responsibilities in this field are still exercised by the Board of Trade and the Minister of Aviation. The Minister of Science is in fact an Overlord – and experience shows that Overlords are seldom, if ever, effective.

The second major omission is the absence of a Scientific and Technical Planning Board. This Board should be the chief instrument to coordinate scientific activity. It should consist largely of scientists and should equal in status the Economic Planning Board with whose work its own should be closely geared. The STPB, serviced by the Minister's secretariat, must help to ensure that the

Government's industrial, economic and social problems take full account of the opportunities opened up by the advance of science, and of the limitations of technical resources. It would largely replace the existing ACSP (Advisory Council on Scientific Policy).

In addition to the existing Councils – the Department of Scientific and Industrial Research, the Medical Research Council and the Agricultural Research Council – there should be created, a Social Science Research Council, and the DSIR, MRC and ARC might well be re-grouped into a Physical Research Council and a Biological Research Council. The Minister and his staff would be charged with reviewing the programmes of scientific work initiated or being executed by all Government agencies, to point out fields of common interest and to stimulate work in neglected areas.

3.7 The Scientific Administration

From Conservative Political Centre, *Science in Industry: the Influence of Government Policy*, a report prepared by a committee under the chairmanship of Mr Robert Carr, MP, November 1962.

The report, which did not become the official party policy advocated a more radical approach than the Conservative government was prepared to concede.

Given the contribution that science must increasingly make to our economic progress, and that the resources of men and money which Britain can devote to research and development are limited, it is evident that we cannot afford to view the nation's defence research and development – accounting as it does in money terms for over half the national scientific effort – as something divorced from our wider economic needs. A collection of decisions involving some £240m. and large numbers of qualified scientists and engineers inevitably affects the outlay which might otherwise go into scientific and technical effort in other directions.

There is not, of course, a straight choice. It is not a matter of machine-guns versus machine-tools, but of balancing the different preferences within the context of an overall view of national needs and resources.

Major decisions about spending on science are already taken at a high level of Government. But (as the existing administrative division implies) they are taken mainly on the basis of arguments between particular Departments and the Treasury – with the outcome depending very largely on the strength of the Department concerned. A long process of consultation between conflicting interests is not the answer. The advantages of expenditure which is basically worthwhile can easily be lost if there are delays in deciding what should be done. What is needed is that all major projects should come to a single decision-making point, and that the decisions should be taken within a coherent framework of policy.

Technical Policy Committee

It is plainly not practicable to subordinate defence research and development to the civil scientific administration. Moreover, an overall view of the nation's research of the kind we have suggested is not something which requires primarily scientific expertise. It involves relating military and industrial research requirements to each other, to the total resources of men and money which are likely to be available, to the relevant demands made by our foreign and defence policies and by economic policy, and to underlying social and political values.

This is a process which can only be effectively carried out at Cabinet level, if it is to have the necessary scope and be backed by the power which is needed to make the results effective. It might be that something of what we have in mind could be secured by making the Minister for Science a member of the Economic Policy and Defence Committees of the Cabinet. But the real need, we suggest, is for an entirely new Technical Policy Committee, specifically charged with the job of formulating the main lines of an overall Government policy on research and development, and seeking to ensure that the whole of Government policy is permeated by scientific and technical considerations. It should have a senior Cabinet Minister as chairman, and its members should obviously include the Minister for Science, the Minister of Defence, and a senior economic Minister.

Such a committee would not of course consider every project for research and development involving Government money. Its job would be to settle major preferences. Within the broad framework both the defence and civil sides must have their own decision-making points, to secure a similar coherence of approach in meeting more limited and detailed needs and in giving effect to the policies laid down.

The civil research administration

Whether or not the resources available for civil research are increased, they will still be limited. We have argued earlier that, within the framework provided by the Technical Policy Committee, both the civil and defence sides need their own 'decision-making' points. For civil research and development it is clear that this must be in the Office of the Minister for Science.

It is neither necessary nor desirable, and probably not even a practical proposition, that the Minister should assume detailed executive responsibility for the work of the AEA and the four research councils – though he should make sure that all five have effective machinery for controlling their own work efficiently, along the lines suggested in the Gibb-Zuckerman Report.* Nor is there any need to change the present fascinatingly crazy set-up of Privy Council Committees, simply for the sake of logic and uniformity.

What *is* necessary is that the Minister should be equipped with the powers and the facilities to set particular research programmes into the context of wider requirements. Limited resources imply selectiveness; selectiveness implies the existence of a coordinating authority with the power to determine preferences in the allocation of resources.

3.8 The Minister for Science

From Federation of British Industries, *Civil Research Policy*, June 1963, pp. 10–11.

The FBI advocated a more powerful, central administrative organisation than the Office of the Minister for Science.

Central civil science : The Minister

(c) We have dealt so far with that part of civil research and development which should fall within departmental votes. This, however, is only part of the story. Basic research in the universities and the whole spectrum of civil work in the various stations under the

* Office of the Minister for Science, *The Management and Control of Research and Development* (HMSO, 1961).

Research Councils and the DSIR as well as the state-supported work in private industry, have also to be considered.

We conceive that the Minister for Science must bear responsibility for determining the allocation of State support as between all these claimants and for the scientific strategy on which it should be based. For this purpose he should be supported by a Council (which should include suitable representation from industry) to assist him in determining the broad lines of policy and, specifically, in the allocation of funds to civil scientific purposes.

(d) Many of the existing agencies covering the administration and execution in individual fields of research are most successful. In agricultural research there may be some room for rationalisation. There is, however, no existing agency capable of administering the support for civil research and development in the industrial field which we are recommending.

The big question of giving general direction to such a programme for industrial research and development must in our opinion be made the responsibility of the Minister for Science advised by a Steering Board, linking representatives of the Government, the universities and industry. There would thus be a two-tier arrangement, the Steering Board receiving an allocation of funds from the upper tier already referred to. Within its allocation, the Steering Board would in effect decide in general terms what to back, giving preference no doubt in areas which appeared to offer the best prospects of industrial growth, or to conform with the concept of 'economic defence'. Its conclusions would determine the *directions* in which an administrative machine under the same Minister would allocate support to approved projects in the field affecting industry.

3.9 Civil Science

From House of Commons Debate, 15 July 1963. Speech by Mr Aubrey Jones, MP, Hansard, vol. 681, col. 89–91.

Jones outlined a proposal for a Ministry of Science and Technology based upon the Ministry of Aviation.

I want now to turn to the question of the central Government organisation for science, because this is, I suppose, the main subject under consideration by Sir Burke Trend's Committee, which has been referred to by my hon. Friend the Parliamentary Secretary. I hope that other speakers will comment on this subject, because it would be nice to think that the House of Commons was, for once, returning to its historic function of playing a part in the shaping of a decision to be made, and not just tramping through the Division Lobbies, 'yeaing' and 'nay-ing' decisions that are, in fact, beyond alteration.

We have, as we know, a Ministry for Science – the emphasis is on the word 'for'. I thought that the hon. Member for Coventry, East, gave quite a fair interpretation of the philosophy behind this concept of the Ministry for Science. That concept is, as I understand it, that the Government should aid, should encourage, should promote science, but that it should play a minimal part, the very slightest part, in determining the direction in which science ought to go. I believe the conclusion to be irresistible that the Government must play a much larger part in determining the direction in which science should go than was conceded four years ago, when the Ministry for Science was established.

I have two reasons for believing that. The first is that, as we have heard this afternoon, it looks as though over the years to come our scientific demands will outrun the scientific resources available. Demand and resources need to be matched, the Government must try to match them and in doing so, inevitably, to a certain extent, determine the direction in which science goes.

Secondly, I think that we have a problem of balance in our scientific effort – a balance between the defence effort and the civil effort, a balance between research and application, a balance between one industry and another. There is only one body to do anything in the way of attempting to secure that balance, and that is Government. For these reasons, I believe that we need much more than a Ministry for Science – aiding, promoting, and encouraging science. We must have a Ministry 'of' Science and since the problem is one of balance, that Ministry of Science should be concerned both with academic science and industrial application – as, indeed, the Department of Scientific and Industrial Research is at this moment.

What should this Ministry of Science be? That is a difficult question. Leaving aside all the details and trying to come to the essentials, we have a choice between two courses. We can have, as now, a very small Department – a 'bus-load of staff' as we have heard it described – and try to strengthen it with a hierarchy of advisory councils. We could strengthen the Lord President of the Council's Advisory Council on Scientific Policy, try to make its composition a little less academic, and have under it, as the FBI report suggested, a steering board for industry.

That is one possibility, and that would be perfectly all right if the problem were only one of apportioning resources between one part of science and another and one industry and another, but the problem is much more than that. The problem is one of improving communication between one part of the scientific spectrum and another; in other words, it is a problem of infusing technology with science and science with technology.

I do not believe that this can be done through the medium of advisory committees which are, after all, collections of private individuals, not served by a working staff, and operating only on policy in its broad generality. This work of infusion, of informing science with technology and technology with science, can be done only at the working level; in other words, it can be done only by a staff, with the advisory councils ancillary to the staff and not the staff ancillary to the advisory councils, which is what we would have if we merely attached one or more advisory bodies to the existing Ministry of Science. We need, then, a staff, and I agree with the hon. Member for Edmonton that there is only one place where the staff is available, and that is in the Ministry of Aviation.

I therefore return to the thought that has been in my mind, which came into my mind some six years ago, and which the passage of the years has not really shaken – that what we really need is a marriage of the scientific functions of the Ministry of Aviation with the scientific functions of the Lord President of the Council, or if hon. Members prefer it, the functions of the Ministry for Science, so that we create a new Ministry of Science and Technology – and I emphasise the word 'technology' – with a wide remit, of which the doing of research and development for the defence Services is only part and, I hope and believe, a diminishing part.

I know that many people will recoil from that suggestion, I am perfectly well aware of all the difficulties, but I have been driven to adhere to this proposal because of the very nature of the problem we face. The problem of science in this country is the bias in the social system. What we did four years ago when we established the Ministry for Science was to defer to that bias and, by deferring to it, we made it much more difficult to escape from it.

3.10 Beyond Research

From Labour Party Policy Statement, 'Labour and the Scientific Revolution', approved at the Annual Conference, Scarborough, October 1963.

Under the Tories, most of the limited scientific help the Government renders to peaceful industry has been distributed among numerous research associations. So long as the unit of production in so many parts of the economy remains small, some dispersion or effort is inevitable.

But a Labour government cannot be content with this. In an important policy statement, the Federation of British Industries has recently admitted that the scale of research required to compete with the industrial giants abroad is often beyond the capacity even of a consortium of British firms. In recommending an extra expenditure of £100 millions on civil research, they have called upon the State for no less than £50 millions.

It is clear therefore that the State itself must be prepared to play a new and major part in civil research and development. The existing National Research Development Corporation, set up by the last Labour government, has already demonstrated the value of direct public initiative in such developments as the hovercraft, digital computers and fuel cells.

In the future we shall build on this experience. The next Labour government will create the new and larger Government organisations required for applying science, on a really massive scale, in civil industry.

But these new organisations must be ready to go beyond the R and D stage. They must be free to engage in production in their own establishments; to sponsor new public enterprise; and to co-operate with private industry in joint ventures on the basis of public participation in the enterprise.

These new initiatives will serve three vital purposes: first, they will stimulate new developments in our older and technologically backward industries such as textile machinery, shipbuilding and machine tools. Second, they will accelerate the development and application of entirely new products and processes. Third, they will recompense the taxpayer adequately for the immense outlay of public funds.

As one of the biggest purchasers in a modern economy – in 1962 Government Departments alone bought over £2,000 million worth of goods and services – the public sector too can play a major role in encouraging industry to introduce new scientific and managerial techniques. Throughout central and local government and the nationalised industries, the Labour government will – wherever possible – make sure that State contracts are used for this purpose. In a wide range of industries, the Government and the nationalised industries as buyers can deliberately demand a standard of design higher than can normally be obtained from industry and can place development contracts deliberately designed to require technological innovation. In the building industry in particular the use of public purchasing power to raise technical standards, to introduce new materials and methods and to encourage necessary standardisation, will yield especially large returns.

The Departments which will place research and development contracts – Education, Works, the Post Office and Transport – will be given the additional scientific staff, of high quality, to ensure that departmental policies take account of modern scientific and technical developments.

3.11 A Minister of Arts and Science

From the Report of the Committee on Higher Education, under the chairmanship of Lord Robbins, Cmnd. 2154, October 1963.

The Robbins Report favoured a governmental administrative merger of arts and science, a proposal not unlike the formula adopted when Lord Hailsham returned to the House of Commons in 1963 (as Mr Quintin Hogg, MP) and assumed the post of Minister for Higher Education and Science.

782. The differences between higher education and education in the schools might suggest that the solution to our problem would be the creation of a ministry that, like the Ministry of Higher and Specialised Secondary Education in the Soviet Union, embraced all the area in which advanced full-time studies take place, from

the universities to the Area Colleges, or indeed all post-school education.

783. At first sight it might seem that these possibilities had much to commend them, in that they would bring all the institutions considered in this Report under one ministerial responsibility. But on closer consideration we do not find them so attractive. If we assume that our recommendations for the organisation of Colleges of Education in university Schools of Education are accepted, then what remains in the non-autonomous sector are the technical and other colleges of further education; and, although we contemplate the eventual transfer of some of these to the autonomous sector, the predominant role of most of the technical colleges is local and regional rather than national, and concerned as much with part-time as with full-time study. Although we believe that the government of individual colleges could be improved on the lines suggested in Chapter XV, we think that administrative convenience and efficiency dictate that they remain where they are in the structure of central and local controls and we see no grounds of principle against it. We shall discuss later a means whereby the advanced work done in these colleges may be related to that in the autonomous institutions.

784. We propose therefore a ministry responsible for a more limited range of institutions of higher education. We recommend the creation of a ministry whose main responsibility is for the autonomous institutions – the institutions that, if our recommendations are accepted, are suitable for control on the grants committee principle. But following the thought developed in paragraph 781 above, we suggest that it would be a most felicitous conjunction if this ministry were also charged with responsibility for other autonomous state-supported activities that are at present administered on principles resembling those of the grants committee. Here we have particularly in mind the Research Councils and the Arts Council; but there are also other bodies designed to forward learning and the arts, such as the Standing Commission on Museums and Galleries, that it might be thought right to bring within the same administrative framework. We of course appreciate that the question of ministerial responsibility for these bodies raises important questions for them and those who depend on them, and we have not attempted to consider these. But from the point of view of higher education, such an arrangement would seem most fitting. In this

way administrative recognition would be given to the essential unity of knowledge; and the nature of the administrative tasks involved would be such as to blend well in a common departmental tradition. Since much of the work would be done through grants committees, the whole would tend to be informed by the special degree of detachment and respect for the autonomy of the institutions and individuals ultimately concerned that is so necessary if the connexion of the State with creative activities is to be a quickening rather than a deadening influence. To emphasise this focus we venture to suggest that the appropriate name for such a ministry would be, not the Ministry of Higher Education, but the Ministry of Arts and Science. It would then naturally assume the present responsibilities of the Lord President in his capacity as Minister for Science.

3.12 Central Administration

From the Report of the Committee of Enquiry into the Organisation of Civil Science, under the chairmanship of Sir Burke Trend, Cmnd. 2171, October 1963.

The setting up of the Trend Committee in March 1962 was the Conservative government's attempt to solve what had by then become a number of pressing administrative problems. There was widespread criticism over what was judged to be a restricted remit (i.e. policy was to be excluded) and over the Committee's findings. Nevertheless these played an important part in subsequent events [cf. 4.3.4, 4.4.4].

We were appointed by the Prime Minister in March, 1962 to consider:

(i) whether any changes are desirable in the existing functions of the various agencies, for which the Minister for Science is responsible, concerned with the formulation of civil scientific policy and the conduct of civil scientific research; and whether any new agencies should be created for these purposes;

(ii) what arrangements should be made for determining, with appropriate scientific advice, the relative importance in the national interest of the claims on the Exchequer for the promotion of civil scientific research in the various fields concerned;

(iii) whether any changes are needed in the existing procedure whereby the agencies concerned are financed and required to account for their expenditure.

4. *Outline of Report.* Our terms of reference are confined to the question of the organisation required in the field of Government-sponsored civil scientific research. We have not examined, therefore, the issues of policy which arise in this field or made any recommendations about either the scale or the content of research programmes. But we have been very conscious that the level of Government expenditure on civil scientific research – which, for the purposes of this Report, we have defined as including development also – has risen rapidly in recent years and that new developments in science and technology will inevitably cause it to increase still further. The Government's organisation for the promotion of such activities must therefore be such as to ensure that the resources provided for this purpose are used to the best advantage, that unnecessary duplication of effort is eliminated and that new 'growing points' are fostered and encouraged.

The Pattern of the Agencies

44. To say that the existing pattern of agencies lacks coherence begs a large question. What is the criterion of coherence? The field with which we are concerned covers the whole spectrum of what is usually called 'research and development'. For purposes of efficient management this spectrum has to be divided into a number of separate parts, which can be assigned to appropriate agencies; and we have interpreted our terms of reference as requiring us to try to devise a logical pattern of such agencies on the hypothesis that each of them should ideally be responsible for a field of activity which is homogeneous in nature and manageable in scope. But at this point a difficulty of classification arises, which is liable to confuse any discussion on this subject. Broadly speaking, there are three ways of sub-dividing the spectrum of research and development – first, in terms of discipline, e.g., chemistry or physics; second, in terms of fields of study, e.g., medicine or agriculture; and, third, in terms of the nature of the activity involved, e.g., pure research or basic research or development. (It was on the last basis that the Committee on the Management and Control of Research and Development – the Zuckerman Committee – identified five categories of this kind: pure basic research, objective basic research, applied

(project) research, applied (operational) research and development.) But the spectrum of research and development is one and continuous. The different branches of subject matter and the various types of activity which it comprises are not completely separate and distinct; any one subject is liable not only to merge imperceptibly into several others but also to involve simultaneously the activities of pure research, applied research and development. In any pattern of agencies that may be devised, therefore, common interests and shared purposes will always seek to transcend the artificial limits of organisational structure; and, in framing our final recommendations, we have sought to bear in mind that the structure should be such as to allow and, indeed, to encourage this to happen.

49. We may summarise our criticism of the existing pattern of agencies by saying that the present arrangements, whereby responsibility for civil scientific research is fragmented and dispersed between a large number of agencies, differing in status, scope and autonomy, are not conducive to either the concentration or the flexibility of effort which are becoming increasingly important in scientific research and development. Science is continually developing and expanding in new directions; and the Government's organisation for the promotion of scientific research should be both sufficiently precise and definite to provide a clear centre of co-ordination for each branch of research activity at any given moment and sufficiently flexible to ensure that new growing points of knowledge, in whatever discipline they may appear, are fostered and that research projects which may have outlived their earlier promise are eliminated. Both precision and flexibility, however, pre-suppose a rational structure, organically articulated.

The Minister for Science

102. If we are right in endorsing the principle of the separate, autonomous agency for the conduct of research and in recommending a reorganisation of the pattern of such agencies in such a way that, in conjunction with the Departments concerned, they will cover the whole field of civil scientific activity, the future Minister for Science must be able to provide effective supervision of their efforts. For this purpose there is one function which he should not assume but several that he should.

103. We take it to be axiomatic that neither the Minister nor his

advisers should seek to exercise supervision over the scientific judgement of the research agencies acting within their respective terms of reference and within the resources which can be made available to them. Their purpose is to enable expert advice, of the highest quality, to be at the service of the Government; and if they were made subject to the supervision of a superior authority, however eminent, which was empowered to pass judgement on their scientific advice, the quality of their membership would inevitably decline and the independence of judgement which is the key to their success would be seriously compromised. We believe that it is essential that this should not occur and that the Minister for Science should maintain the long tradition whereby he and his predecessors have not intervened in the exercise by the Research Councils of scientific judgement within their own fields.

104. On the other hand the Minister should be given wider powers to enable him to play a more positive and effective part than hitherto in promoting scientific research and development. First, he should seek to encourage conditions in which research can flourish and its results can be put rapidly to use. This involves, among other things, the provision of qualified scientific manpower in numbers and types sufficient to meet national needs.

105. Second, he should assume, and discharge to Parliament, the responsibility for assessing the financial requirements of the agencies, for settling with the Treasury the funds to be made available for the promotion of civil scientific research and development and for allocating these funds as grants-in-aid among the agencies. As we have explained in Part II the Minister of Science has not hitherto been directly responsible for the financing of the research agencies, apart from the Atomic Energy Authority and NIRNS. The Treasury has had to decide between the claims of the research councils; and, where these have exceeded the available resources, there has been no means of deciding which items could be regarded as marginal on scientific grounds and could, if necessary, be eliminated. The different channels through which the Councils have received their funds have added to the difficulty. This system may have sufficed so long as the total outlay of the research councils and other agencies was small. But today the total expenditure of the research councils alone is of the order of £35 million; a single new item of expenditure may amount to £1 million or more; and major facilities for research in nuclear physics or for space research may

cost many millions. In such circumstances it is necessary that the Minister for Science, in recommending such expenditure to his colleagues, should be fortified by a full assessment of the scientific case for each project involved in relation to the needs not of one, but of all, the agencies for which he is responsible.

107. Third, the Minister should assume general responsibility for keeping the whole field of civil scientific research and development under continuous and comprehensive review, in order to ensure that deficiencies are made good, that unnecessary duplication is avoided and that the broad deployment of effort over the different areas of the field is in accordance with the national interest. On these matters the Minister would have the advice of the Advisory Council which we recommend in paragraph 112; he would work with, and through, the various agencies; and, while he would not intervene in their exercise of their own scientific judgement, he would decide priorities in cases where a major commitment of resources was involved. He should also keep under continuous review the machinery required for the promotion of civil scientific research sponsored by the Government, in order to ensure that, whenever necessary, adjustments are made in the distribution of functions and responsibilities between the various agencies concerned, that new organisations are created when necessary and that bodies which have outlived their usefulness are terminated.

109. A Minister for Science, with substantive functions on the lines which we have described, would naturally need to maintain close contact with his colleagues in the Government, particularly those whose Departmental functions include responsibility for some measure of research; and we assume that the necessary arrangements for this purpose would be made as part of the normal operation of the machinery of government. It would therefore follow that the Privy Council Committees should be dissolved and that their place should be taken by the Minister himself, who would inherit the power to appoint the governing bodies of the research councils (and would assume a corresponding power in relation to the new agencies), together with the power to issue formal directions to all these bodies. In exercising his power of appointment the Minister would no doubt take advice from appropriate sources, including the Royal Society; while, as regards the

powers of direction, he would presumably use these very rarely and would never seek to call in question the independence of the governing bodies in matters of scientific judgement.

The Advisory Council

112. We recommend that a new advisory body to the Minister for Science should now be constituted. This would be akin to, but in important respects different from, the existing Advisory Council on Scientific Policy. Its terms of reference might be the same as those of the present Council (see paragraph 41); but they would be invested with new significance and authority, corresponding to the enlarged functions which we have proposed for the Minister for Science. The Council would therefore be responsible for advising the minister on all aspects of his duties, as summarised in paragraphs 104–8, i.e. on scientific manpower; on the allocation of available resources between the agencies; on national scientific needs as a whole, including the fostering of new 'growing points' and the elimination of duplication of effort; on international scientific policy; and on the organisation of the administrative machinery for the promotion of scientific research and development.

113. In order that the Advisory Council may be fitted for these tasks, we consider that it should consist entirely of independent members and that the Chairmen of the research agencies, or other official representatives, should be invited to attend, as necessary, only as assessors. We propose that the Council should consist, in addition to an independent Chairman, of not more than fourteen persons, half of whom should be scientists. The membership should include three or four industrialists, at least one economist and several individuals with wide experience of public affairs. They should be appointed by the Minister for Science, who would no doubt take advice from the appropriate sources before nominating members; they should serve on a part-time basis, although they should be prepared to devote a substantial amount of time to the Council's affairs; and their period of service should not normally exceed five years. The Council would be advisory to the minister; and all decisions based on its advice would be decisions of the minister. The minister might well publish an annual White Paper on Scientific Policy and Expenditure and, on occasion, reports by the Council on particular topics arising from its work.

The Office of the Minister for Science

116. The reorganisation which we have proposed will impose additional functions and responsibilities on the Office of the Minister for Science. If the Advisory Council is to provide the minister with sound and authoritative advice, it will need to be serviced by a considerably strengthened secretariat, which should comprise technically qualified as well as purely administrative staff and should be provided by the Office of the Minister. In addition, the Office will need to cooperate closely with the various autonomous agencies concerned with research; and it would be advisable that, for this purpose, a standing committee of the Chairmen of these agencies should be established, over which the Permanent Head of the Office would preside. This Committee should also include, as necessary, representatives of Departments with substantial interests in research and development. It would be concerned with management rather than policy; and its functions would be to ensure effective service for the Advisory Council, to discuss problems of common concern to the agencies and, in general, to take the initiative in ensuring that the administrative structure as a whole remains alert and efficient. We have emphasised in Part III that any workable division of the unitary field of science must inevitably leave some untidiness at the lines of demarcation, whether the distinction is drawn in terms of different types of subject matter or different administrative agencies. It should be the responsibility of the inter-agency committee to secure the coordination required to reduce to the minimum the duplication or omission of effort in such cases. Equally, many of the problems of modern research and development require an interdisciplinary approach in so far as they involve several agencies simultaneously; and this tendency will be liable to grow as programmes of scientific research and development assume an increasingly international character. Here, aslo, the Committee will have an essential part to play in providing administrative coordination for the agencies concerned; and for this purpose it may be appropriate to include in its membership, as occasion requires, representatives of other bodies, such as the UGC and the Royal Society, particularly in the light of the part which the Royal Society will undoubtedly continue to play in promoting non-Governmental international scientific cooperation.

3.13 Cultures Apart

From *The Spectator*, 8 November 1963, p. 584.

Comment on the Trend Report.

Anyone in the Government who really thought that Mr Wilson's ambitious talk about Labour and Science was going to be countered by the publication of the Trend Report must now be a disappointed man. The Report's tampering with the arrangements for nature conservancy and oceanography and its scheme to dismantle the Department of Scientific and Industrial Research and then set it up again under another name can hardly have disturbed the opposition one little bit. Yet it was not so much the Trend Committee at fault as the exceptionally limited terms of reference. The Committee were asked to report on the existing organisation of scientific research and its relationship to government. It cannot therefore have taken them long to reach their conclusion that the organisation has at present no coherent or articulated pattern. What is so surprising is that they were then content to go on to produce only plans for a rationalisation of the present structure. The Report's recommendations, if accepted, will bring us into line with the plans outlined by the Haldane Report of 1918.

The Report has placed the usual muddled stress on the need for research to remain independent. In doing so it has lined itself up with Robbins in ignoring the fact that some loss of independence (or, which is what it means, the introduction of an order of priorities) is not by any means tantamount to loss of academic freedom. In America the President's scientific adviser, Dr Wiesner, can call for information from any government research institution he pleases, and it is significant that he is just initiating an investigation into science and government which may result in throwing over for good the system where a non-scientist Government gives the scientists everything they want in the name of science, progress, independence and ignorance.

The Trend Committee has failed to criticise two contradictory trends in British government practice. The one is this overwhelming stress on independent research. The other is the need felt by certain Ministries to appoint bodies like Neddy and the Schools Council to do their research for them. The former is a principle of government by good luck (which may well turn out to be bad luck);

the latter is an attempt to govern by consent and consultation, where decisions wait upon adequate and, in the widest sense, scientific research. There is little doubt which is the more satisfactory. We ought now to be moving into the position where around the seat of government there is the equivalent of a postgraduate university, and there is indeed no good reason why it shouldn't be called a Government University. It would take over many of the functions of the Civil Service and of the existing research institutions. It would cover not only the obvious subjects like nuclear physics and aeronautics but also sociology, criminology and economics, none of which either the Government or the Trend Committee seem to regard as being in the slightest way scientific. One of its main advantages would be that it would be a sure way of reuniting brains and government, or at least brains and the kind of knowledge they ought to command.

3.14 Scientific Policy and Manpower

From House of Lords Debate, 11 March 1964. Speech by Lord Bowden, Hansard, vol. 256, col. 482–4.

A view on the Trend Report.

I think the most important criticism that can be made of the Trend Report is that when one discusses it among practising scientists, as I have had occasion to do many times during the last few weeks, most of them dismiss the greater part of its deliberations as fundamentally unimportant; even, one might say, irrelevant. This may seem a harsh thing to say, but I think the truth of the matter is that the really difficult problems with which the Trend Committee might have been concerned were excluded either by its very narrow terms of reference or by the way in which it chose to interpret them. It was in the first place charged to consider only the science with which the Ministry is concerned. It rather deliberately added development itself; but then, in the course of its deliberations, it almost entirely ignored this aspect. Furthermore, it ignored the very complex problems which are posed by the Ministry of Defence, by the Atomic Energy Authority and by those large research establishments controlled by the nationalised industries; and in some important respects it even ignored the research of the universities. It is hardly possible to write a Report about a matter so important if one ignores something like four-fifths of the field with which the Report should perhaps have concerned itself.

My own criticism is that the Committee should have gone back and asked for different terms of reference. It should have concerned itself, in fact, with the total range of research that is being done in this country; because, to isolate from the rest the part with which it was concerned makes production of a sensible plan almost impossible. The effect of the operations of the Ministry of Defence, of the Ministry of Aviation and of the Atomic Energy Authority on the whole field of science is quite fundamental and cannot be ignored. The other grave omission from the Report is any sense that the object of applied research and industrial research is not merely intellectual satisfaction, but the production of designs which can be manufactured and sold at a profit. This fundamental point seems to have been almost completely ignored by the Trend Report. This particular ultimate objective, the prosperity of British industry and the achievement of better designs which are manufacturable at a profit, should have been, to my mind, the primary concern to any Committee concerned with research and development. The fact that this was not mentioned seems sufficient condemnation at least of the terms of reference with which the Committee was burdened.

3.15 Science and Politicians

From *The Economist*, 22 August 1964, pp. 701–2.

Comment on possible changes in science administration likely to follow the 1964 general election. The journal favoured the Aubrey Jones proposal of basing change upon the Ministry of Aviation.

Whichever government returns to power after the election will find itself faced, among other things, with the need to decide how many of the exclusively science-based industries Britain can afford. Both political parties, and Whitehall, are singularly ill-equipped to do this.

The solutions currently being discussed are not particularly apt either. The Conservative government has killed its own palsied little creation, the Office of the Minister for Science, and made it part of the Ministry of Education with Mr Hogg as the minister. The intention, if the Conservatives are returned after the election, is to set up special research councils – for pure science, for industrial development and for natural resources – along the lines which have worked successfully enough in the past for medicine and agriculture. They would be financed by the Ministry of Education and would in their turn finance individual research projects. All very neat and tidy – but with the basic underlying absurdity that the ministry which has to deal with primary schools, and teachers' training, ought not to be asked also to concern itself with industrial research, with the direction of atomic energy and with whether the Government should put up £20 million for the development of a computer that knits Fair-Isle patterns. One should, perhaps, be grateful for the fact that Mr Hogg did not offer also to absorb the Ministry of Aviation into his education empire; for the aviation ministry spends rather more than half the amount that the Government sets aside for research. It has charge not only of the aircraft and associated industries, but is also responsible for one of the key science-based industries in the country – electronics.

It would be reasonable, after Mr Wilson's brave words of a year ago, and his cautiously deliberate accumulation of scientific advisers, to expect Labour to offer something better. But the signs suggest that it will not. At present such unscientific procedures as listening among the grass roots are the only means of learning what Labour has in mind for science, but the message from the grass roots is not encouraging. It suggests that Labour would be prepared to carry the split between so-called 'pure' and so-called 'applied' science one step further than the Conservatives and share them between separate ministries. Pure science would stay inside the Ministry of Education as a sort of tame pekinese for the universities. Applied science and industrial development would be removed to form the nucleus of a Ministry of Technology.

Why? What is wrong in the 1960s with having one Ministry of Science that would be basically responsible for seeing that the Government's £450 million or so annually was spent in the right places? The fact that this might have been wrong several years ago when science was still fuzzy round the edges does not mean it is still wrong now, when it has become generally recognised that scientific advisers, however eminent and however sound their advice, are not much use to modern government unless they have some kind of permanent base and a permanent office. It may seem incongruous to complain that the way funds are allotted by the Government to science in this country is fossilised in tradition, but how otherwise can one explain the astonishing inconsistencies in the way the money gets shared out? On this year's budgeting, £240 million of the Government's £400 to £450 million will be spent on defence research, in the aircraft industry and on electronics – i.e. will pass through the hands of the Ministry of Aviation. Atomic energy will get, from one source or another, rather more than £100 million. Help to industry, in the form of contributions to the industrial research associations, will not even reach the microscopic sum of £3 million. What, other than tradition, can justify such an extraordinarily lop-sided allocation of resources?

3.16 End of the Advisory Council on Scientific Policy

From the Annual Report of the Advisory Council on Scientific Policy 1963–64, Cmnd. 2538, December 1964.

The ACSP summed up its own work.

11. The main issue, and the most enduring issue, which has exercised the Council during its seventeen years of existence has been the question of the scale and balance of our national civil scientific effort. We cannot pretend to have done more than achieve a measure of the magnitude of the problem, and of the great responsibilities which would rest on those who, unlike ourselves, might one day be charged with the task of cutting up the cake of our national scientific resources.

12. We have had to form judgements about the scale of the scientific effort of other leading nations. The importance of this task has increased progressively both with the widening realisation that the success of our whole economy depends on the wise exploitation of modern science and technology, and with an increasing awareness of the fact that a major reason for our deteriorating position in world markets is the success with which other countries have exploited their own scientific efforts. We have had to compare the scale of our efforts in technology and applied science with that in fundamental research; we have considered the balance of effort between the various sciences, and in some scientific fields, of the relationship of our national to our international efforts. Matters such as these, and the emergence of vastly expensive areas of scientific activity – for example, nuclear physics, oceanography, and

space technology – were the major reason why it became necessary to devise a new machinery for reviewing civil scientific requirements as a whole.

15. This is the most important issue which we pass on to our successors. The problem of priorities in science and technology lies at the heart of national science policy, and therefore of our national destiny.

3.17 Role of the Council for Scientific Policy

From the Council for Scientific Policy, Report on Science Policy, Cmnd. 3007, May 1966.

The CSP was created by the Labour government to replace the former ACSP; in spite of its name its brief was narrower: to advise the Secretary of State for Education and Science in his responsibility for the research councils. The government also set up an Advisory Council for Technology to assist the new Minister of Technology; unlike the CSP this committee never reported publicly.

Functions of the Council for Scientific Policy
4. The Science and Technology Act 1965 created a fundamentally different position for this Council as compared with its predecessor, the Advisory Council on Scientific Policy. Whereas the mandate for our predecessors ranged over the whole of science and technology, they advised a Minister who had general oversight of a wide range of research establishments (he appointed the members of Research Councils and in principle could give them directions) but few major direct financial powers. The new Act provides for the Secretary of State for Education and Science directly to finance the Research Councils; but at the same time the greater part of the Government research establishments oriented primarily towards technology and its application in industry have become the separate responsibility of the Minister of Technology.

5. Our terms of reference are to advise you in the exercise of your responsibilities for civil science policy. The range of problems has therefore become a narrower but much more intensive one; instead of advising on a very wide range of problems arising largely *ad hoc* and concerning frequently the organisations and spending powers

of other Ministers, we have now to advise you in your statutory responsibility for determining the overall pattern of the resources of the research councils. The Councils now cover every field of basic and applied science, within and without the universities, including nuclear physics, astronomy and oceanography which formerly were partially excluded, and in addition important fields of 'mission oriented' research, as in medicine and agriculture. We regard this specific and novel responsibility as our prime task, and the attainment of a solution the first challenge to the new organisation.

6. There are of course wider aspects. We must also take account of developments in the universities and other institutions of higher education, of defence research, and of industrial research and technology. There are also the important parameters imposed by the supply of qualified manpower. We have as members or assessors representatives who are experienced in all these fields. In the end a synthesis of all these issues must be achieved, but this will involve the separate responsibilities of a number of Ministers. Our first task is to demonstrate within the field for which you are directly responsible how criteria for the development of science can be formulated and applied in practice.

The Scope of Science Policy

7. It is necessary at the outset to deal with the misconception that the advance of scientific knowledge itself can be directed from the centre. This would be to misunderstand the original and spontaneous nature of science. The advance of scientific knowledge cannot solely be achieved by the arbitrary selection of national scientific goals and by committing resources of men and money to them. Because science is original it is also unpredictable: neither the provenance of a new idea nor its ultimate applications can reliably be foreseen by scientific policy-makers. At the other end of the spectrum, however, are technological goals involving directed research and the better application of existing knowledge to specific missions. When the scientific principles are foreseen it is possible to mount massive mission-oriented attacks upon problems as in the case of the development of antibiotics, nuclear energy, and space communications. Such programmes often have enormous economic or social implications, and have great value in stimulating, in order to reach a defined goal, a concerted attack on a broad range of

scientific problems. But they need to be carefully managed in relation to their objectives if they are not to result in more and more elaborate solutions to problems which may have changed fundamentally. While the majority of mission-oriented research is not within your responsibility, important examples (such as Agricultural Research) are; and we intend therefore to devote attention to the issues posed by them.

8. Science policy does not therefore direct the advance of scientific knowledge, though it may well be concerned to encourage or to direct the application of the results of scientific advances. The tasks of science policy are of another kind: to maintain the environment necessary for scientific discovery; to ensure the provision of a sufficient share of the total national resources; to ensure that there is balance between fields and that others are not avoidably neglected; to provide opportunities for inter-fertilisation between fields, and between the scientific programmes of nations. These responsibilities are distinct from the assessment of scientific merit or the management of research, which are responsibilities appropriate to Universities or Research Councils as the case may be, and we depend upon them for performing these functions. Our task is to look ahead on a national and indeed an international scale; to take account of the relationships between science and other national activities all of which may have a competing claim for resources, and to provide the basis for national acceptance of the scale of resources which are needed for the healthy evolution of science.

3.18 Administrative Coordination

From the Organisation for Economic Cooperation and Development, *Reviews of National Science Policy: United Kingdom and Germany*, Paris, 1967, pp. 142–5.

Extracts from the confrontation meeting of experts held on 2–3 March 1966.

For the British Delegation, Sir Frank Turnbull acknowledged that the question of coordination between the Department of Education and Science and the Ministry of Technology was one he and his colleagues were frequently confronted with, and that the organisation might indeed arouse the fear of a loss in policy coherence. All he could say, however, was that this fear was not justified in practice. He referred to the various institutionalised links between the two Ministries and their advisory bodies, but put equal stress in the importance of the 'Cabinet system' for coordination of policy, the detailed working of which could not be disclosed, and on the great experience that the civil servants in these and other Ministries had in continuous inter-departmental consultations and contacts as part of their normal work.

In the general discussion, Professor Rexed (Sweden) inquired about the function of the Science Adviser to the Cabinet, whose position was indicated in the Diagram showing the United Kingdom organisation but not discussed in the report. The leader of the United Kingdom Delegation – referring to the origin of that office during the war – made it clear that this function was not to be regarded as the top of a pyramid of the various advisory organisms for civil science. The Scientific Adviser to the Cabinet participated when matters of a scientific nature came before the cabinet, or its Committees. This question arose in the context of a general debate on the structure of the advisory system. Dr Hocker had asked the United Kingdom Delegation whether, especially in view of the division of the main responsibility between two Ministries, it would not be desirable to have an advisory body in an independent position that could consider the affairs of both with impartiality and authority. He was supported by Mr Beckler (USA), who referred to the system in this country and suggested differentiating between an overall function of 'looking across the board', and another that would serve the direct operational purposes and interests of the individual department. As an example for the functions of the former, he cited the importance of the interactions between military and civil research that needed to be integrated at the top level of government.

In his replies, Sir Frank Turnbull made it quite clear that the United Kingdom system did not contain an independent high level body for advice and policy planning over the whole field of science. In the United Kingdom system of Government, Ministers each had defined fields of responsibility, and there were now scientific advisory bodies in the respective fields of defence, technology, and education and science, advising the appropriate Ministers. After talking about the various links between the Council for Scientific Policy and the Advisory Council for Technology, he went on:

'This provides a close interconnection between these two Ministries. I do not think myself that it would help the working of this system to contemplate the amalgamation of the two Councils. They advise different Ministers about different things and are composed of different kinds of people. The idea of a sort of high Council for Science, which is adjacent to the Prime Minister or President, has its attractions, of course, but the question does arise, at any rate in our system, whether if you had such a body it would be asked to advise on anything except the broadest major issues. We believe you need scientific advice integrated into departmental operations ... On the whole, we believe that in our system of Government it is best to have scientific advice tendered directly to the Minister who is responsible for a given field. I think the existence of this kind of system is one of the reasons why scientists feel they are properly brought in . . .'

The question how the coordination of political and scientific responsibility was organised and functioned was viewed rather differently for each country by the Examiners. The basic impression was that the science side had a larger amount of independent policy initiative in Germany, whereas in the United Kingdom it had been brought more closely into the governmental structure. This was regarded as involving in Germany the risk that the Government share in general science policy was rather of the 'responsive' than of a truly initiative character, in the United Kingdom the risk that the main scientific bodies might identify themselves too closely with the respective departmental interests and become remote from the scientific community.

In replying to the same question for the United Kingdom Delegation, Sir Frank Turnbull said that the words 'division of responsibility' in the question did not seem appropriate, since the real problem should rather be seen as one of 'how to integrate the scientific point of view and the political point of view in reaching any decision on this sort of subject matter', and then outlined the important function of scientists within the respective department. On a later occasion, he referred to the question whether the scientific community felt their interests satisfactorily represented at Government level. He first pointed to the long history of the

Research Council system and the important role of scientists within the Government structure during and after the war.

'A great many of the leading scientists today in the United Kingdom have had this experience and they do, therefore, have an understanding of the way in which the administrative and financial machinery of the Government operates and what you can reasonably expect to do and what you cannot . . . I would think myself that this is one of the most important factors in this situation . . . I was asked whether the scientists felt now that *vis-à-vis* other parts of the Government, and in particular the Treasury, their case was adequately represented. I think there was a period when some doubt was felt about this, . . . latterly I think there has perhaps been some feeling that because we have four Research Councils, the interests of basic science are divided and have not been presented as a unity to the Government in a coordinated way. The main purpose of constituting the new Council for Scientific Policy is to meet this feeling. It is also to meet a need felt on the administrative and financial side, namely that it is very difficult to deal with four bodies, some of which have interests which are very near to one another. On the whole I think the creation of this Council has met the concern for the time being. We had, of course, the Advisory Council on Scientific Policy earlier, but it had no financial functions at all, and the essential difference between the former Council and the new one is that the latter now has an advisory function in relation to finance.'

3.19 The Central Advisory Council for Science and Technology

From Parliamentary Questions, 25 October 1966. Question by Mr Neil Marten, MP, Hansard, vol. 734, col. 821–2.

Mr Marten asked the Prime Minister what steps he will now take to further the Government's policy of scientific advance.

The Prime Minister (Mr Harold Wilson): As the hon. Member knows, the Government are concerned not only to promote the growth of our scientific and technological resources at a rate which is sensible in relation to our needs and resources, but also to ensure that they are deployed to the best national advantage. In addition to strengthening interdepartmental coordination for this purpose, we propose to establish a small central Advisory Committee, under the chairmanship of the Chief Scientific Adviser to the Government, comprising not only representatives of the Council of Scientific Policy, the Advisory Council on Technology and the Royal Society, but also some other independent members.

Mr Marten: In view of the serious rise in the brain drain of scientists and technologists, will this Committee deal with the subject of the brain drain? Secondly, does not the Prime Minister believe that if we had a sensible national space programme, for example, that would retain in this country a number of our space scientists to work in Britain on the frontiers of knowledge?

The Prime Minister: The function of this Committee will be to help to advise the Government on the broadest strategy of allocating resources among some of the fundamental types of research. The hon. Gentleman mentioned space, which is important and highly expensive, and there are others which may be less expensive but perhaps no less important. On manpower and the position of scientists, the increase in the provision for scientists is helping to provide work for a much more variegated group of scientists than before. This is a question for the Advisory Council on Scientific Manpower.

Mr Dalyell: In view of the widespread opinion that British science tends to be far too pure, would the Prime Minister make certain that the six Government Departments involved give urgent and coordinated attention to the recommendations of the Swann Committee and, in particular, to the ideas of the Swann Committee on post-graduate work inside industry?

The Prime Minister: Yes, Sir, but the Ministry of Technology was created particularly to expand that part of our scientific effort which means clothing the results of pure science and technological discoveries with 'know-how' for the use of industry, and a considerable part of the increased expenditure in the last year or two has been on the practical rather than the pure side.

Mr Hogg: How does the new body which the right hon. Gentleman has announced tie in with the existing functions of the Council on Scientific Policy, which up to this moment has been performing precisely this role? Why does the right hon. Gentleman continually cloak his failure in this department by inventing new gimmicks?

The Prime Minister: For the simple reason, which the right hon.

and learned Gentleman, who has enough knowledge to appreciate, will understand, that the Council for Scientific Policy deals with pure research, with the work of the research councils for which the right hon. and learned Gentleman was once responsible. But there is also the whole field of Government expenditure in technological research and the application of science to industry. It is therefore necessary to bring together not only these two, but the Royal Society, which is not a Government body but which also has important functions, so that the whole scientific strategy of the nation can be reviewed by one body. When the right hon. and learned Gentleman speaks of failures, he will recognise the size of the increase in provision for scientific and technological research since he ceased to hold office.

3.20 Science and Technology: Departmental Responsibilities

From House of Lords Debate, 14 June 1967. Speech by Lord Shackleton, Minister without Portfolio, Hansard, vol. 283, col. 945-7.

Reasons for the divisions between the Department of Education and Science and the Ministry of Technology.

Let me give some of the reasons for the division between the Department of Education and Science and the Ministry of Technology. Again, I am speculating to some extent in regard to this matter. There may be arguments which sound smooth rationalisation, but nevertheless there are sound reasons for dividing for administrative purposes the support of the scientific environment from the support of the exploitation of science. I hope that your Lordships will bear with me in speculating in this way. These two operations are separate in objectives and even in methods, though there may be overlapping in countless directions. For the former, the discovery of new knowledge is the supreme factor; for the latter, the attainment of results measured in terms of economic performance. Each of these tasks is difficult and complex. One can go on for some time exploiting existing knowledge with considerable success, but unless also resources are preserved for adding to that knowledge, the time comes when one is forced to buy new knowledge from one's competitors and one is deserted by scientists of the calibre needed to keep abreast of world science. I know there are those who will argue the contrary to this, but I submit that there is validity in it.

Each of these jobs is man-sized and needs in the present context the weight of a full departmental organisation. The Ministry of Technology are concerned with getting the fruits of research for the economic system. If the Department of Education and Science were to undertake this task also, they would become an economic as well as an educational and scientific organisation. This would be too much to ask of them. Again, it is too early at this stage to return to the point of view expressed by the noble Earl about the division between higher education and schools. On this, we may find something more when Dr Dainton reports his finding. On the other hand, the Ministry of Technology does not need to be responsible for the Research Councils in order to foster the modernisation of industry, any more than it needs to be responsible for schools and universities in order to get the right distribution of skills in the economy.

I would say to the noble Earl, and I am sure that he will agree with me in this, that we can have too big and too centralised an organisation, as many large industrial concerns are aware. Super-Ministries have not had a happy history and they can fail simply because the burden of policy-making and management ought to be shared out more. It may be argued that the ideal state existed when Mr Hogg (then Viscount Hailsham) and the noble Earl were responsible for the Ministry of Science, when the Minister had the satisfaction of being able to say that he could get the whole of his Ministry into a single bus. This is an ideal state for an organisation of that kind, but unfortunately, as things developed, it was not possible to confine it to one bus. It has now reached the stage, if we are to make the progress which is necessary, and on which the scientific community are supporting the Government and the country so well, when it is better to have two small organisations which know what they have to do rather than a larger body whose right and left hands are not always able to be coordinated. It is extremely important to see that the right hand does not cut off the left hand when it happens to be looking in another direction.

Let me turn to the position of the universities in this context. The universities are concerned not only with education. They are in an indissoluble way concerned also with research. They must occupy a key position between education and research, and in our view it is sound that one Minister – again, I put these views with humility – should be able to take a view over the whole of this territory. The University Grants Committee and the universities ought therefore to be linked with the Department that is concerned with education. I suggest that the present division between the Department of Education and Science and the Ministry of Technology is a wise experimental solution and is to be welcomed as giving to each organisation a clear run at objectives which in some respects are limited. No horse can run well if it is facing too many ways at one time.

There is a difference of ideology between the scientific environment, where knowledge overrides time, place and country, and the competitive environment of technology, where results in time and within the budget are what count. Of course, there is much science in technological issues and there are vital scientific elements in all the main issues of State, whether in foreign affairs, industry, defence, transport, or any other. But scientific issues there are not necessarily overriding, whereas, in my view, they must be so in the environment of basic science.

What counts in this complex and difficult field is that the interface between basic science and its application should be understood, or at least recognised, and that sufficient numbers of scientists and technologists of the highest abilities should recognise in the application of science and technology at least as worthwhile a challenge as the discovery of new knowledge itself; and lastly, that there are in practice as well as on paper effective arrangements for consultation and coordination.

PART TWO
Practical Problems

4 Research and Development

In this chapter we examine the government's involvement with the performance of research and development activities. This takes several forms: for instance, the government is directly responsible for the R & D performed by its own employees in the laboratories of the research councils and the departments; secondly, there is the work performed cooperatively for industry – with government financial aid – by the research associations; and thirdly, there is the work performed by teams or individuals in universities, industry or institutions and supported by government grants mediated by the University Grants Committee, the research councils, the departments and bodies such as the National Research Development Corporation.

The arrangement of this chapter follows broadly upon this division by sectors; thus the extracts cover:

1. Problems of choice and priority with which the government is faced.
2. Government research establishments.
3. Research councils.
4. Government support for research in industry.

Much of the Introduction was devoted to a description of the administrative developments whereby this spectrum of activities grew. This is not repeated here; in keeping with the character of the other chapters in this part of the book, we shall illustrate certain key themes and developments.

We have already described the historical accident which led to the devising of the (then unique) administrative system for the DSIR. The Haldane Committee on the Machinery of Government [cf. 1.5.1] noticed the features of this which have since been constantly used in its justification and as the basis for the administration of the research councils of more recent foundation – and so successfully that the Treasury Committee of Civil Research could write in 1928 as if there were something divinely ordained about the British research council network [4.3.1].

High among these characteristics has been the relative freedom from detailed government interference accorded to scientists working for the councils [4.3.3]; but a consequence of this, not foreseen by Haldane, was an apparent detachment of the councils from the true concern of government, the welfare of the community – a detachment partly shared by the government's own research establishments [4.2.1, 4.4.2]. But it must be conceded that to demonstrate the value of research is difficult enough a problem now and it was obviously even more so when the understanding of research evaluation was at a more primitive stage [4.4.3].

The ending of the second world war and the consequential escalation of research budgets (from a total expenditure by government on civil science in 1939–40 of £3.6 million to one of £42 million by 1961–2) has provoked a more questioning attitude to the role and work of the research councils and the government establishments, and a quarter of a century later one can see that the same questions are still relevant: what is the proper function of the state's own establishments [4.2.2, 4.2.5]? Under what circumstances should one be closed down, or, alternatively, passed over to, say, a university? To what extent should basic research be a responsibility of national establishments as distinct from its traditional home in the universities [4.1.3]? To what extent is it desirable or feasible to foster collaboration between state establishments and the universities [4.2.4]? Is it possible to determine the optimal disposition of state funds for research, and is it possible to detect 'gaps' in the spectrum of research activities which might be filled by an injection of funds [4.1.2, 4.1.4]? Is the Haldane principle of research council independence still valid today? To what extent are the research councils justified in using the power given to them by virtue of the funds at their disposal to build up 'centres of excellence' or to concentrate on certain areas of research [4.1.5]? How might the defence research establishments be integrated in the civil research network [4.4.6]?*

The asking of questions such as these and in particular concern over the rising costs but less certain benefits of research led to two major investigations into government research policy and performance. The first was begun in 1958 'to inquire into the techniques employed by Government Departments and other bodies wholly financed by the Exchequer for the management and control of research and development' and resulted in the Gibb-Zuckerman

* In 1967–8 the total government expenditure on R & D was £545.8 million, of which defence R & D accounted for £257.6 million.

Report of 1961 [4.2.3, 6.7]; and the second some four years later to inquire generally into the organisation of the whole of civil science, which led to the publication of the Trend Report of 1963 [3.12, 4.3.4, 4.4.4].

With the benefit of hindsight, the Trend Report must be seen as the more significant of these two inquiries and was the first indication that the research council formula was not indefinitely elastic. Trend advocated the setting up of two new research councils on lines similar to the existing ones and the breaking up of DSIR – part to form one of the new research councils [4.3.4] and part to go into a new Industrial Research and Development Authority [4.4.4]. The latter was to be an autonomous body controlling some former DSIR stations, the research associations and the National Research Development Corporation (NRDC) with a view to achieving greater coordination of research for industry.

That such a proposal was made is indicative of the current feeling that industrial research in the country was in general unbalanced – three-quarters of it being concentrated in three sectors (aviation, the chemical industry and electrical and electronic engineering). But it is clear that any government's ability to influence policies and actions in private industry is relatively restricted. The NRDC, designed to provide capital to aid the exploitation of inventions, is one postwar innovation [4.4.9]; a second one was the granting of civil development contracts – civil equivalents of military development contracts – but overall these cannot be judged to have been a success.

The Ministry of Technology, set up by the Labour government in 1965, took over most of the functions allocated by Trend to the still-born IRDA and accreted many more during its five-year life. But it became, as we noted in the Introduction, much more directly involved economically and financially with industry and more specifically with its restructuring, for instance, through the operations of its satellite the Industrial Reorganisation Corporation; the former Minister, Mr Anthony Wedgwood Benn, expressed it thus in his Cantor Lecture to the Royal Society of Arts (18 November 1968): 'The matters that have absorbed most of our time, and which I personally have been most interested in, have been the problems of industrial structure, because if you want commercial success you have got to get the structure right first.'

The research association scheme was devised under the pressure of industrial shortages in the first world war [4.4.7], and although yielding very tangible benefits to industry, has never succeeded in becoming self-sustaining. The primary worry in the interwar years was the ensuring of adequate financial support for all research activities, not only the research associations [4.4.3]. By and large this has not been the case since the war, but more subtle problems have come under scrutiny: what are the appropriate functions of a research association? and in what circumstances is one most effective [4.4.8]? A more sceptical attitude to the government's own research establishments has had inevitable repercussions for the research associations; their role was left conspicuously undefined by the proposal in 1970 to set up a British Research and Development Corporation (BRDC) [4.4.5, 4.4.6].

As the 1970s open there is some measure of consensus between the two main political parties as to the future of the R & D establishments: that the establishments ought to earn their keep, and that their resources and facilities ought to be directed to a greater extent towards the private sector. One reflection of this new attitude was embodied in the proposals contained in the Green Paper *A Framework for Government Research and Development,** published at the end of 1971. This contained two reports (by Lord Rothschild and Sir Frederick Dainton) designed to make the research councils more accountable to government. The Green Paper and the controversy it sparked off were the subject of Parliamentary investigation by the Select Committee on Science and Technology in the spring of 1972.

4.1 PROBLEMS OF CHOICE AND PRIORITY

4.1.1 Science, Industry and Society

From *Nature*, 27 March 1937, vol. 139, pp. 525–7.

An early glimmering that a national scientific stock-taking – virtually a commonplace since 1945 – was becoming a necessity for the United Kingdom.

What is clear is that we have reached a stage when it is imperative to take stock of the nation's resources and facilities for research, both academic and industrial. To neglect or starve research is to invite national peril, but the danger of unbalanced development in certain fields while others remain scarcely explored is too great to be ignored. Some means of securing a better distribution of the national effort in research is indeed long overdue, and such distribution must take due account of all research activities and facilities,

* Cmnd. 4814, November 1971.

whether by the Departments of State, at the universities or within industry. To such a national stocktaking the Department of Scientific and Industrial Research might well make a decisive and indispensable contribution. Few organisations are in a better position to encourage the prosecution within industry itself of intensive research in the applied and physical sciences, with all the important reactions which such research continually has on the technique and outlook of the underlying sciences themselves. The relations between academic and industrial research in the physical sciences at the present time, the extent to which progress depends on team work, independent of whether the investigation is prosecuted within the university or within industry, the inspiration and assistance which academic and industrial research in these fields are continually bringing to one another, give fundamental research in the physical sciences special claims upon the support of industry, apart from the rapidity with which discoveries in this field are applied to industrial purposes. If the main burden of research in such fields were accepted by industrial resources, it should be possible to redistribute research facilities at the universities so as to endow much more liberally that research in the social sciences which is so urgently needed if we are to attain an understanding of the problems of society, its organisation and adaptation to the forces playing upon it today, whether within industry or outside it. Without such understanding there is dire peril that the riches with which scientific knowledge could endow us even now may never be enjoyed.

4.1.2 The Balance of Scientific Effort

From the Annual Report of the Advisory Council on Scientific Policy 1959–60, Cmnd. 1167, October 1960, pp. 13–14.

The ACSP set its face against the possibility of devising a rigid plan for science; at best it could plan only a broad strategy but in the last resort it is the individual who counts. Later in the same report the ACSP turned to a postwar problem which really had no prewar analogue: the escalating costs of equipment and installations for advanced research; nuclear physics was, and still is, a prime example. In this passage the ACSP spells out some of the consequences of this economic necessity. In the final paragraph the Council looks further ahead.

3. If such a review is to have a chance of being effective, it is necessary to define clearly both its aim and its limitations. Science and technology are not static; and the concept of balance is not easy to apply. Balance implies equilibrium; and disequilibrium is of the essence of scientific progress. It would be futile to seek a formal structure which could hold good in the face of constant change in these fields. Any rigid plan would be harmful as it would soon be out of date. Similarly, 'gaps' in science and technology cannot, in our view, be filled in any conventional sense. As the limits of knowledge are forced outwards, the area of contact with the unknown is apt to increase: to fill an apparent gap may merely be to provide a point of vantage from which many more are visible. Even the definition of a gap is debatable: are 'gaps' simply neglected subjects or should the term be restricted to fields where lack of work is serious for economic or other reasons?

4. At the same time, we recognise the prime importance of devising criteria by which the points can be identified at which additional effort and resources will pay greatest dividends. It ought to be possible, on the basis of a comprehensive review, to indicate broad fields where further work is needed; or where a new assessment or re-orientation of effort seems desirable. We are concerned not with administrative matters, but with the strategy of scientific affairs. We believe that the Research Councils and other governmental agencies operating in scientific and technological matters are the appropriate bodies to interpret our recommendations in terms of programmes of work.

5. By our reference to strategy we must not be taken as implying that the advance of scientific knowledge can be directed: but from time to time obstacles can be removed, encouragement and resources can be provided for people who have ideas, and the best use made of our limited resources. Basic advances in science depend upon the creative individual. This process cannot be controlled; and theories that such progress can be produced merely by pouring in money and resources are without foundation.

38. We have already referred in paragraph 12 to the high cost of some projects in science. This high cost, with the attendant problems of staffing and organisation, calls for serious consideration especially in both universities and industry. In the past, new fields of scientific study have developed in universities round brilliant

individuals, who not only made notable contributions to the advancement of their fields of study, but in due course sent out their pupils and disciples to create and develop similar studies in other universities. This procedure has been highly satisfactory in the past where scientific research was relatively inexpensive, but when the total cost of activity on a worthwhile scale may run into millions of pounds the situation is quite different. Our resources are limited and more concentration of our effort is required; attempts to pursue certain types of scientific activity in many different universities can result in a dispersal and dilution of effort such that in the end we achieve much less than could reasonably be expected from the total expenditure involved. There is, moreover, the possibility that large installations of equipment, which must be serviced by many specialised technicians and which has a high obsolescence rate, may distort the pattern of a university and lead to a situation in which the direction of its research may be determined, not by the scientific stature of men available, but by the machines in existence.

39. We already have a clear example of these dangers in the field of nuclear physics; we feel sure that a greater return would have been obtained for our expenditure in this field had there been less dispersal of effort in the early stages of its post-war development. The situation has since been recognised by the creation of a National Institute for Research in Nuclear Science (NIRNS) where will be located the very large and costly machines needed for nuclear physics research which will be used largely by university workers. We also note with satisfaction recent developments in the regionalisation of electronic computers and in proposals for joint operation of nuclear reactors; such developments have been smoothly arranged by adaptation of our existing machinery for the financing of research. But the creation of a National Institute such as the NIRNS is a new departure, and the extension of this idea may call for some adjustments of our scientific organisation. These are important matters; for not only in the fields mentioned but in others (such as for example optical and radio astronomy and space research) it is necessary to ask at what level of expenditure, both in resources and man-power, we are prepared to operate, and how we should avoid dispersal of effort and unnecessary duplication of highly expensive research facilities. We hesitate to name a precise figure, but certainly any project which involves expenditure on a large scale might best be located in a central institute, available to

workers in universities and similar institutions. Alternatively, such activities should be confined to one or two universities at most. We are aware that this implies some sacrifice of the freedom of universities to make their own arrangements; but we see no alternative if a country of our size is to participate effectively in these expensive fields of research. The form of organisation and control appropriate to National Institutes of this kind needs to be further considered.

40. The principle can be carried a stage further; there must equally be a point at which the cost of adequate participation is so great as not to justify mounting an entire independent programme in this country at all; and where international cooperation provides the only appropriate way of continuing in these fields without subordinating the entire scientific effort of the United Kingdom to a few major programmes. We believe that in space research, nuclear physics and oceanography we are approaching this stage; and that some form of international cooperation (as has been brilliantly demonstrated, for example, in CERN) will in the end be unavoidable.

4.1.3 Basic Research and Universities

From the Organisation for Economic Cooperation and Development, *Science and Policy : the Implications of Science and Technology for National and International Affairs*, Paris, 1963, pp. 10–11.

The economic appraisal of research became a major concern of the 1960s; such considerations ran the risk that purely speculative research might be downgraded as unproductive. In this extract the OECD spelled out the 'seedcorn' view of basic research as essentially a university preserve of primarily educational significance.

37. There is always the danger that exclusive policy attention to urgent near-term problems will result in failure to maintain adequate financial and institutional support of basic research activity, particularly since practical utility and policy relevance are more remote here than in any other part of scientific endeavour. No failure could be better calculated to bring all scientific advance to a stop within a very few years. Pending development of satisfactory means of measuring the return on investment in basic research, it should be a fundamental tenet of every nation's policy for science to devote a generous proportion of its total scientific resources to the support of balanced, continuing, healthy basic research effort. No

precise percentage figure can be applicable in all cases, so that it becomes a primary concern of policy thinking to arrive at the appropriate sum for each nation. Errors in this calculation should be allowed only on the generous side.

38. The increased volume and enhanced practical utility of research can also threaten the traditional role of the university as the sanctuary of basic research. Some research scientists are over-burdened by administrative and contract-monitoring responsibilities. Others find too heavy a teaching burden an obstacle to creative research. Large and expensive research equipment requiring elaborate engineering maintenance can induce imbalances in departmental strengths, hiring procedures, and pay scales. Yet the university remains the proper home of basic science because of its free environment, the stimulus of young minds, and the seminal influence of inter-disciplinary contact. A principal objective of national science policy must be to insure that university admin-istrations, budgets, laboratories, and standards of excellence are maintained at the highest levels possible.

4.1.4 The Case against the European 300 GeV Accelerator

From Department of Education and Science, *The Proposed 300 GeV Accelerator*, Cmnd. 3503, January 1968; the Minority Report by Sir Ewart Jones and Sir Ronald Nyholm, p. 55.

The European Organisation for Nuclear Research, usually known as CERN (the acronym derived from its French name) proposed in the mid-1960s a powerful particle accelerator as a successor to its 28 GeV machine. This was to be supported internationally at a cost put at 1,776 million Swiss francs. The high proportion of this cost allocated to the United Kingdom led the Secretary of State for Education and Science (Mr Patrick Gordon Walker) to call for advice from the Council for Scientific Policy and the Science Research Council. Both bodies supported, albeit with some reserva-tions, the project, but Jones and Nyholm did not and their view is given here. The case is unusual in that the government of the day published all the advice they had been given even though they decided on economic grounds not to participate, a decision reversed in 1970.

We recognise that a good case on purely scientific grounds has been made for proceeding with this project. But we are also acutely aware that there are other scientific fields which, cultivated and nurtured as nuclear physics has been in recent years, would yield still richer harvests. Some of these are of no less scientific interest and have much greater potentialities for benefiting mankind. In several of these fields we are leading the world, we should be able to offer generous support where such ability is demonstrated, but this we cannot afford to do.

Our participation in this project means that for the next ten years the Science Research Council expenditure on a single branch of physics will continue to consume more than 40 per cent of the funds which seem likely to be available. Thus the Council's intention gradually to reduce the proportion of its resources going to nuclear physics will be completely frustrated. We are convinced that to continue to spend such substantial sums in this direction is not in the national interest and that many scientists in this country, if they are properly appraised of the situation, will view the prospect with dismay. Can we expect them calmly to accept a slowing down of their activities whilst we go ahead with a huge project which will directly benefit only two or three hundred academic physicists?

We cannot reconcile embarking upon this massive new commitment with our present unsatisfactory and worsening situation. Can high energy nuclear physics justly claim to be of such overwhelming merit and importance? We should urgently be seeking opportunities of investing comparable, and if possible larger, sums in projects which offer some prospect of material advantage to the community and which at the same time serve to train useful scientists and technologists.

4.1.5 Selectivity and Concentration in Research

From Science Research Council, *Selectivity and Concentration in Support of Research*, 1970, pp. 9–10.

The policy developed in the late 1960s, by which research councils' spending was restricted, necessitated a more critical attitude towards the support the councils were able to give. The SRC's approach was set out in this pamphlet; an extract from the summary is given.

The principles which will be followed in adopting a policy of greater concentration and selectivity are broadly as follows:

1 Certain areas, within a discipline or embracing a number of disciplines, will be selected for more favourable than average support during a given period, on the basis of a review of their special potential for advancing basic science, or their economic or community value, or all three. Other important criteria will be the economy of scarce manpower and the optimum utilisation of unique or expensive facilities in universities, national and international laboratories and in industry.

2 A limited number of university departments will be given more favourable than average support to enable them to concentrate effort in selected areas; such departments will be selected on the basis of their leadership, past achievement, present expertise, or other relevant factors (e.g. ability to collaborate with industry).

3 This concentration of resources will be planned by shifting to favoured areas from less-favoured areas rather than by simple addition.

4 Nevertheless it will be an essential part of SRC policy – and well publicised – that some support will always be available to any outstanding individual in any part of any subject for work of sufficient 'timeliness and promise' (e.g. imagination, novelty or relevance to valuable aim).

5 The pattern of preferred topics and places will be kept under continuous review and not frozen. This, with 4 above, will make it possible for any department or individual to grow, with SRC help, from a small start to a major group in any field, provided there are sufficient ideas, effort and backing from the university itself. With a limited growth rate for SRC as a whole it will be necessary to reduce support in major areas where programmes have been completed or have lost their impetus in order to provide backing for new centres.

6 The degree of concentration, i.e. the proportion of the funds given to selected areas or to selected departments, must depend upon the nature of the subject (e.g. need for very large equipment), the existing degree of concentration, the resources available (e.g. the number of trained experts in the subject) and so on. But it will be the subject of appropriate review by SRC, in the light of open discussion with university and other people concerned.

7 Some of the principles to be followed in exercising selectivity in support of astronomy, space and nuclear physics research differ in

important respects from those arising at present in other branches of science and engineering. Because of high threshold costs and large capital installations consideration has to be given to the creation of regional or national facilities or participation in international organisations. The selection and support of university teams by the SRC to take advantage of such facilities requires close collaboration between university workers and the staff of national and international laboratories and the obligation to accept the discipline which such collaboration entails. Similar considerations are likely to arise in other fields where major engineering installations are required to be shared by universities in furtherance of particular research programmes.

4.2 GOVERNMENT RESEARCH ESTABLISHMENTS

4.2.1 Government Research is Uncommercial

From J. D. Bernal, *The Social Function of Science* (Routledge, London, 1939), p. 148.

For political as well as economic reasons there is also on the part of Governments an extreme reluctance to take any active part in research on the application of science. If a Governmental laboratory arrives at any result which could have commercial value, it is not in a position to exploit it, rather it is definitely prevented from either selling the process to an industrial firm or operating it on its own account. The general principle is laid down that in no circumstances, outside military requirements in war time, should Government Departments compete in production with industrial enterprise. The inevitable result is that the attitude of Government institutions towards applications of research is almost entirely negative. They have no incentive towards the extension of applications, and they consequently tend to concern themselves with answering specific demands of industries, particularly in dealing with those cases where it is only a matter of finding the remedy for some recognised difficulty in industrial production. Thus Governmental scientific research is not, outside the Soviet Union, capable of providing either the impetus towards the new application of science or the rational control and direction of such applications as there are.

4.2.2 Aims and Functions of Government Research Stations

From the Report of the Department of Scientific and Industrial Research for the Year 1959, Cmnd. 1049. May 1960, pp. 11–12.

We have completed our survey of present and projected research at the Department's research stations and have made no new changes this year in organisation or scope of work. However, our review had led us to reconsider the broad aims and functions of the stations as a whole and to redefine their terms of reference clearly and in relation to modern needs. The new definitions have been sent to Directors to help them guide the expansion of their activities along sound lines.

Basic research in the stations, we have said, must be directed to the general advancement of technology and applied science. We recognise however, that each station should have the freedom to follow up new ideas that fall within its field of interest, and so to include in its programme some research which is not directed towards any immediate need of industry or Government. In considering their programmes stations must take into account both the ability and responsibility of industry and research associations for organising applied research, and also the scope of related work in universities and colleges of technology.

Each station has before it a limited number of clearly defined objectives, chosen so as to yield, if attained, the maximum national advantage from the resources available for research. Provided that these objectives are practical, the work may consist of basic research, applied research or development: the criterion for selection is not the nature of the work but the end to which it is directed. Projects in the research programmes are also limited in number, so that proper progress can be made on each of them.

Directors aim to ensure that the appropriate Government interests are linked as closely as possible with the stations' work and, in consultation with research associations where appropriate, they aim to secure financial or other cooperation from industry. We are aware that industry attaches importance to the fact that the Department's stations are completely unbiased both in the selection of research projects and in the presentation of results.

Within these terms of reference the major purposes of the research stations are to:

(a) keep their fields of research under constant review in order to define objectives and help the Government, industry and the public to maintain a lively interest in the value of research;

(b) conduct research which can provide information to central and local government on matters such as air and water pollution, road safety, noise and the extinction and prevention of fires, in which the Government has a clear responsibility for protecting the health, safety and welfare of the citizen;

(c) carry out research and development in subjects, such as the natural resources of the country and the design and construction of buildings and roads, which are important to the Government and which affect the efficiency of industry as a whole;

(d) extend the frontiers of knowledge in applied science so that industry can be provided with the basic information required for the solution of particular problems;

(e) pay special regard to the research needs of industries that lack an adequate scientific background, and to research problems that are common to more than one industry;

(f) carry out particular researches, in cooperation with industry wherever possible, which will enable the stations to appreciate industry's problems more fully and to recognise those fields in which more basic research is most urgently needed;

(g) provide for industry national and international standards of measurement of various fundamental physical quantities (such as length, mass and time), related secondary standards and reference materials;

(h) conduct research on matters of broad public interest; and

(i) disseminate the results of research and secure their application.

4.2.3 Research Management

From the Office of the Minister for Science, *The Management and Control of Research and Development*, 1961.

The report, generally known as the Gibb-Zuckerman report after the names of the successive chairmen of the committee which wrote it, was one of the first attempts to grapple with the problems of cost control in civil and defence research and development.

1. We were set up in May 1958 by the Lord President of the Council, whose responsibility for the oversight of Government science later became that of the Minister for Science. Our terms of reference were:

'To inquire into the techniques employed by Government Departments and other bodies wholly financed by the Exchequer for the management and control of research and development carried out by them or on their behalf, and to make recommendations.'

4. Our terms of reference focus on the techniques of management and control. From this we concluded that we were not expected to question directly the size or content of research and development programmes or the existing division of responsibility between the various organisations involved. But starting from a consideration of techniques of management, we have felt free to follow any argument to its conclusion even if that were to involve the consideration of some major change in, say, the inter-Service coordination of defence projects, or the structure of the Scientific Civil Service.

81. In our view, pure basic research is best carried out in the environment of a university rather than in that of a Government research establishment. It is a characteristic of universities that they provide their members with the necessary freedom to pursue any line of inquiry they wish to follow and, broadly speaking, at whatever pace their inclinations dictate. In the choice of work a research worker in a university need take no account of the needs of industry or of national priorities. Normally he is free from the immediate pressures exerted by those concerned with the applications of the results of his work. Colleagues who work in his own and related disciplines provide a vital stimulus and can act as touchstones for new ideas. We do not, of course, suggest that comparable conditions cannot be established outside the academic world, but they are far less likely to be realised either in Government or industrial research establishments, the justification for whose existence is that they bring science to bear on the solution of relatively immediate practical problems.

83. Unlike what we call pure basic research, objective basic research, undertaken in order to try to fill a known gap in a field of

potential practical importance, is very much the direct concern of Government research organisations. It is, of course, also the concern of industrial laboratories and, to a certain extent, of the universities. Its great importance to Government science does not, however, mean that Government organisations should arrange for their own establishments to undertake, either wholly or partly, all such research.

105. It is often said that British scientists are good at basic research and bad at developing their discoveries. Penicillin and radar (and even nuclear energy) are frequently used as illustrations of this generalisation. Other examples, no doubt, could easily be found. The charge is also made that we do not spend enough money on applied research or development, and that we do not attract enough of our best men to these aspects of scientific endeavour. This is partly due, so we are told, to the success of the Royal Society and the universities in building up the prestige of pure science, and partly to the failure of many employers – particularly those whose firms depend on the empirical development of craft techniques – to understand the part that applied science can and must play if their enterprises are to survive.

106. There is some truth in all these generalisations. The importance of applied research and development is certainly not widely enough appreciated, and we believe that the skills they demand tend to be underrated. While it is true that applied science usually breaks less new ground than does basic research, the qualities – personal as well as scientific – needed to make a success of an important piece of applied research or development are not less estimable, and no less rare, than are those which characterise the higher flights of basic research. Of many important and stimulating examples of applied research carried out in Government research organisations we would mention the development by the Safety in Mines Research Establishment of 'foam plugs' as a fire fighting technique; the work on diffraction gratings at the National Physical Laboratory which has enabled the National Engineering Laboratory, in collaboration with industry, to improve the precision of control of machine tools; the production at Rothamsted Experimental Station of virus-free plants of considerable practical application, e.g. a virus-free strain of King Edward potato with a yield 10 per cent higher than the normal stock; and the opening up by

the Post Office research staff of the possibilities of long-distance transmission using circular waveguides, whereby a large number of signals can be transmitted simultaneously through a guide of only a few inches in diameter.

115. The more varied responsibilities of DSIR make their problem of management far more complicated. The Council of DSIR forms its views of user requirements on the expressed or assumed needs of industry for the kind of information on which industry will base its own applied research and development work; on the research requirements which relate to public services such as roads and buildings; on the needs of the community generally in relation to such matters as air and water pollution, road safety, noise and fire control; and on the specific need of Government Departments for information which may affect administrative decisions. Proposals to open up some new field of research usually come from Directors of stations and their staffs, or from their advisory boards and committees, less frequently from the Council and its committees and headquarters staff. Proposals for new projects within an already established field of work come mainly from Directors and their staffs and advisory committees.

116. The pattern and scale of effort having been agreed by the Council, the Director of a station is given a large measure of freedom to determine the details of his programme and to support or stop subsidiary items of research. At some stations, Advisory Boards with a substantial membership of scientists and industrialists assist the Directors in formulating the general programme. In addition, there are close contacts between DSIR and its stations on the one hand, and industrial research associations on the other. As an experiment, the Council set up about two years ago small Steering Committees to bear the responsibility for the programmes of certain of its stations where special circumstances existed. These Steering Committees consist of a member of the Council, specialists from outside DSIR, the Director of the station concerned, and other representatives of DSIR. The member of Council or the Deputy Secretary of DSIR is usually the Chairman.

117. We understand that the purpose of these arrangements is to bring the user or potential user into contact with all levels of the organisation, from the laboratory bench to the Council itself. Potential users should therefore have the opportunity of playing

their part in determining both the details of the programme of individual stations and the broad pattern of DSIR's work.

4.2.4 Advantages of Collaboration between Universities and Government Establishments

From Council for Scientific Policy, Report of the Working Party on Liaison between Universities and Government Research Establishments, Cmnd. 3222, March 1967, pp. 2–4.

The Working Party was set up in April 1965 by the Council for Scientific Policy and under the chairmanship of Sir Gordon Sutherland, FRS, 'to consider the question of liaison between Universities and Government establishments for educational purposes, especially in the research field'. In the extract the value of this type of contact is sketched out.

10. The Government-financed organisations surveyed cover a considerable range of activities and purposes. First there are the laboratories set up by Government Departments to pursue research in a field which is the direct responsibility of that Department, e.g. Defence or Technology. There are also Research Council units and institutions, set up in or near Universities, with the object of fostering research, often of an essentially fundamental nature, and there are laboratories, such as the Daresbury and Rutherford high-energy nuclear physics laboratories, administered by the Science Research Council, to provide central research facilities for Universities. Within such a wide range of laboratories, there can be no uniformity in the desirable types of linkage with the Universities, which also differ widely in their special interests and functions. Nevertheless, some general observations can be made concerning benefits which closer collaboration would bring to the country as a whole, to the Government laboratories and to the Universities.

11. From a national point of view, closer collaboration between Universities and Government research establishments is desirable for several reasons. Joint use of expensive capital equipment in certain fields of research will achieve a desirable economy in the use of public money, without denying the proper supply of research facilities to all concerned. More important, however, closer collaboration is desirable in order that the facilities for post-gradu-

ate work that are now available in Government research establishments, both in respect of equipment and potential supervising staff, should be more fully utilised. Closer collaboration will lead to a greater appreciation in the Universities of the contributions which the academic scientist can make to the solution of vital national problems and to industry, while the establishments will realise more quickly the practical significance of advances in pure science in many of their problems. Pure science is, of course, not the preserve of the Universities, any more than applied science is confined to Government and industrial establishments. Indeed, it is just because the line between pure and applied is frequently quite ill-defined, that more interaction is desirable. Some Government laboratories have found it necessary to build up large research groups in 'objective basic research' because the Universities were ignoring particular fields. This has subsequently created difficulties for Universities endeavouring to enter those areas. Sometimes, also, a University may have already been working in a particular field, but perforce on a scale insufficient to satisfy the national need. If a Government establishment then enters this field, the University may have to abandon it because it finds that the major results are being more quickly obtained by the larger effort that the establishment can deploy. Closer collaboration and mutual appreciation of one another's problems could prevent this happening in the future.

12. For the establishments, one of the most obvious advantages lies in the recruitment of junior staff from research students who have learnt of the work of the establishment from university staff or from carrying out part of their thesis work there. Since the discipline of giving a systematic course of lectures on one's general field of research is often a valuable stimulus to research, some of the staff in the research establishments would undoubtedly benefit from devoting a part of their time to advanced teaching in a University. One of the factors which makes university research laboratories often livelier than Government research laboratories is the continuous influx of fresh young minds. Each year there is the stimulus that about 30 per cent of the youngest age-group is replaced by newcomers, whereas Government laboratories, apart from those going through a rapid phase of expansion, recruit annually a very limited number of new young staff. There is always some danger that establishments may become ingrown from the lack of fresh stimuli

of this kind. A wider policy of giving fixed-term fellowships, or short-term appointments, to a fraction of the younger scientists employed at Government laboratories, and a more frequent involvement of a few of the permanent staff in the teaching and research training of the next generation of scientists, could be very beneficial in this respect.

13. For the Universities, the establishments are the repository of much scientific and technological knowledge which could be of considerable value in advanced teaching and research training. Some research establishments are recognised national centres of excellence in their chosen fields of research, and university departments with similar interests have much to gain from having closer relationships with them, especially through easier access to specialised and expensive research facilities. By using selected staff (on a part-time basis) from research establishments, the Universities can be relieved of the need to recruit new full-time staff to give instruction in certain specialised topics, especially in the early stages of creating a new sub-department. One of the functions of a University is to give the best possible training to the scientists and engineers who will subsequently work in industry. Here, some of them will be engaged on very large projects, initially as members, and later as leaders of a research and development team. Certain Government laboratories can often provide much better facilities and a more suitable environment for a large part of the training of such men in research than the Universities.

4.2.5 Finding a Role for the Government Research Establishments

From House of Lords Debate, 14 June 1967. Speech by Lord Todd, FRS, Hansard, vol. 283, col. 962–4.

I have always held that there are only two sets of circumstances in which research can be relied on to be pursued successfully over a long period. One is when the research is directed to clearcut-economic, and preferably changing, objectives. This is the sort of situation which obtains in industry. There you can carry out research for a long period with a more or less permanent staff. The other way in which you can do it, and where you do not need to have economic objectives, is where you give effect to it by the simple device of having it depend on a continuous through-put of transient young research workers who do not remain in the group doing research for any long period – where, indeed, there is a training function involved for the small permanent staff and the whole thing is kept alive by the continued flow of young and fresh minds. That is the sort of thing that obtains only in the universities.

The trouble is that Government establishments which do not have a clear-cut economic objective (and there are many such) operating as they do with a largely permanent staff, do not fall into either of these categories. As a result, they tend to be ineffective, except perhaps in the initial phase of their development when the staff is new, young and enthusiastic. The desire expressed by those members of the staff from the Government research establish-ments to teach and to have Ph.D. students is, to my mind, merely a manifestation of this situation.

Some Government establishments we must of course have. There are those which concern what I may call in a general way public services: the Road Research Laboratory, laboratories deal-ing with standards, like the National Physical Laboratory and certain others. But, for the rest, apart from those dealing with specialised defence problems, I think that, in the main, these Government establishments should be more directly associated in some cases with universities and in other cases with industry.

I greatly welcome moves which have recently been made to develop the links between some establishments and the universi-ties: for example, between the National Engineering Research Laboratory and the University of Strathclyde, of which I have the honour to be Chancellor. I think this is a good type of move. But I consider that the greatly desired increase in contact between the universities and industry is not going to be achieved by merely appointing a few industrial or Government scientists and techno-logists as honorary professors, setting up one or two joint projects and allowing a few young men to take external Ph.Ds. You need something much more direct than this if you are to get any-where.

Why cannot we be bold for once and, for example, take the National Engineering Research Laboratory right out of the Ministry of Technology and put it slap under the control of the University of Strathclyde? Not only do I believe that this would be more effective so far as research is concerned, but I believe that if this kind of thing were done the development of applied science in universities and, what is more, the recruitment of their graduates into industry would be greatly simplified. After all, this kind of thing has been quite successful in the United States, and I do not see why it should not be successful here.

If you turn to the other side, I do not see much future for places like Harwell or the National Gas Turbine Research Establishment if they carry on as Government research establishments. Why should not the National Gas Turbine Research Establishment be put slap in the aero-engine industry? After all, that is where it really belongs and where it ought to have been developed in the first place. I find it rather hard to see the need for a Government station to provide research requirements for modern industry. If there is a need for this kind of thing, then we should stimulate the industry concerned to put its house in order. If we did, I think we should hear a great deal less about 'brain drains' and the difficulty of recruiting young scientists into industry.

I know that I can be accused of prejudice when I say this because I am a chemist. But I would remind your Lordships that the chemical industry does not complain about or suffer so much as other industries in regard to these particular things. For instance, if you look at the findings in the Report of Professor Swann's Working Party you will see that the chemical industry recruits a large proportion of people from the universities, many more than most other sciences. If ever there was a research-based industry it is the chemical industry.

This makes me think. I wonder whether any of these other industries in trouble, and perhaps the Government, too, should not look at what has happened in the chemical industry, an industry where, so far as I am aware, they have never shown either the need or the desire for Government research stations to aid them, but in effect have done their own research and cooperated fully with universities all the way through. Action in the Government research establishments along lines of the kind I have just mentioned will certainly present administrative difficulties, but I do not think they are insurmountable; far from it. Certainly they should not be allowed to prevent action, because I believe that the idea of Government research establishments in areas outside the public services may have been a good one in its day, but that day has passed.

4.3 RESEARCH COUNCILS

4.3.1 Justification for the Research Council Network

From H.M. Treasury Committee of Civil Research, Report of the Research Coordination Subcommittee, 1928, pp. 10–11.

24. In its support of utilitarian research the Government has been led to set up, as we have seen, a system consisting mainly of a tripartite organisation, viz.

(a) for agricultural research which deals with the production and protection of plant and animal life needed for human use. With this must be associated fisheries and forestry research, and we shall deal later, in each case, with the form the association takes, and should take;

(b) for medical research, which deals by no means only with the cure of disease. It deals with the proper development and right use of the human body in all conditions of activity and environment, as well as with its protection from disease and accident, and its repair;

(c) for industrial research, which deals with the materials and methods used in all forms of manufacturing industry.

25. It will be seen that all the material fields of human need and activity are covered, or may be covered, by this three-fold system. There will be overlapping, often useful and desirable, at the common boundaries, and it is not difficult to find subjects which for convenience, based on historical accident or technical convenience, may be left outside this triple scheme. Of these, one instance is astronomy and its applications to navigation, and for this our needs far back in history led to early special provision.

27. If it be agreed that the organisations for utilitarian State-aided research are inclusive enough, it may nevertheless be asked why they should be three-fold and not merged in a single scientific organisation. In our own view, it is fortunate that the history of events, if not any conscious Parliamentary intention, has led to the differentiation of three organisations and not to their coalescence. We express this opinion from the point of view of their scientific efficiency and not here from that of their constitutional positions in relation to the Government and Parliament. We believe that a

single scientific organisation would have been unwieldy and that while the optimum size and intricacy may be approached by each alone, it would be exceeded if fusion came about of two or more of the group. It may be agreed further that as each organisation has grown, or is still to grow, the development of this new kind of State machinery proceeds in large part by trial and error. Each developing along its own path is more likely so to reach the model most suited to its own work, while each, at this point, or that, may gain from noting the procedure of its companion organisations and may give similar opportunities to them for sympathetic but critical attention. The most conclusive argument, however, for mutual independence among the three appears to us to lie in this, that each of the organisations in its daily work of fostering all the stages of research between primary discovery and the final applications of its results to the practical business of the professions, the trades and other interests, has to deal with a distinctive world, distinctive in its history, its psychology and its conditions of work. All must deal with scientific men; but the agricultural organisation must have expert and sympathetic knowledge of farmers and of the veterinary profession. That for medicine has to deal with a particular profession, ancient and highly organised, which has special responsibilities towards human life. That for industrial research is concerned with men in industrial and commercial life. These differences seem to point to the necessity for separate organisations, separately named, however closely they may be linked for their mutual help, and make it difficult to argue from the detailed practice of one to that of the others.

4.3.2 British Work on the Atomic Bomb

From the Report of the Advisory Council for Scientific and Industrial Research for 1947–48, and a Review of the Ten Year Year Period 1938–48, Cmd. 7761, August 1949.

The Department of Scientific and Industrial Research was maintained throughout the second world war and demonstrated a high degree of flexibility, though much of its research effort was switched into the defence and the civil defence field; the Road Research Laboratory, for instance, turned to the civil engineering problems of civil defence. However, two projects taken up by groups within the DSIR are particularly noteworthy: radar and the atomic bomb. The Department's work on the initial, development stages of the latter is described.

3.2.1 As is well known, after the observation of nuclear fission by Hahn and Strassman in December 1938, and its subsequent verification and explanation by Frisch and Meitner, almost every nuclear physics laboratory in the world started on work related to this phenomenon. The British universities, notably Birmingham, Cambridge, Liverpool, and London were well to the fore. The possibility of an explosive release of energy had always been realised, and it was natural that attention should be turned to this work on the outbreak of war. Early in 1940, a committee was set up, first under the Air Ministry and later under the Ministry of Aircraft Production, to coordinate the work on uranium and to report on its possibilities. The work was on a small scale; but in the summer of 1941 the Committee was able to report that the indications were that a weapon was possible and that methods existed to produce the material; the Committee therefore recommended that the project should be carried out on a large scale with the highest priority. This report was endorsed by the Scientific Advisory Committee. The War Cabinet accepted the recommendation, and responsibility for the work was transferred from MAP to DSIR since, at this stage, the problem was chiefly one of fundamental development of theoretical physics. The story of the British part in the development of the atomic bomb has been told and only the outline of the story need be repeated here.

3.2.2 The Prime Minister asked the Lord President of the Council, at that time Sir John Anderson, to be personally responsible for the development, and the transfer of administrative responsibility became effective on 1 December 1941. A new section of the Department was set up to deal with this investigation, which, for security reasons, was given the name 'Tube Alloys Research'. ICI Ltd, agreed to release Mr W. A. (now Sir Wallace) Akers to become Director of Tube Alloys Research. To advise him on the project, the Lord President set up and presided over the Tube Alloys Consultative Council, consisting of the Chairman of the Scientific Advisory Committee of the War Cabinet (Lord Hankey and later Mr R. A. Butler), the President of the Royal Society (Sir Henry Dale), the Secretary of the Department of Scientific and Industrial Research (Sir Edward Appleton) and Lord Cherwell.

The Director of Tube Alloys was advised by a Technical Committee, under his own chairmanship; the members of this Committee were Professor (now Sir James) Chadwick, Professor R. Peierls, Dr H. Halban, Dr F. E. Simon, and Dr R. E. Slade, joined later by Sir Charles Darwin and Professor (now Sir John) Cockcroft, Professor M. L. Oliphant and Professor N. Feather.

3.2.3 The first task, on the administrative side, was to make arrangements with various universities and industrial firms for work to proceed at a greatly increased rate. During the first year, work was located at the Universities of Birmingham, Cambridge, Liverpool, and Oxford, and with ICI Ltd, and Metropolitan-Vickers Ltd. Other contracts were placed with other industrial firms on specialised aspects of the work. Special arrangements were made to preserve the rights to the Government in patents resulting from inventions in the work.

3.2.4 Towards the end of 1942, it was decided that if some of the work were to proceed on the necessary scale, it was unwise for it to be in the UK. Arrangements were therefore made with the Canadian Government for a joint effort in Canada, and during December, 1942, and January, 1943, a team of some thirty-five British scientists was transferred to Canada.

3.2.5 It became increasingly evident during 1943 that the Allied cause would best be served by USA, Britain and Canada pooling their resources on the uranium project, and the necessary arrangements were made. The American effort was already very large and facilities in USA for this work were clearly better. It was therefore decided that the most rapid progress would be made by concentrating the research in USA and placing British scientists at the disposal of the American authorities. From this point onwards, therefore, the emphasis was on transferring and recruiting British personnel for work in USA and Canada. Certain fundamental work continued in the UK, and work continued on the British gaseous diffusion plant design; but the practical work in this country necessarily slackened considerably. Professor Chadwick moved to Washington in 1943 as Chief Scientific Adviser in USA, reporting direct to Sir John Anderson. He was followed by a number of British scientists including Professors Massey, Oliphant, and Peierls who assisted various aspects of the USA project, and by Professor Niels Bohr of Denmark, who came to this country to help the project. For the rest of 1944, the main task was to recruit and supply specialised staff for the USA and Canadian projects, until there were 50 British scientists on the Department's strength in USA, and about 120, at the peak, in Canada.

3.2.6 The story of the American project is now well known and is told in the general account of the development of methods of using atomic energy for military purposes, which has been published.

3.2.7 By the spring of 1945, the American project was very near its successful completion; most of the fundamental problems had been solved, and British scientists began to return to the United Kingdom.

3.2.8 By this time, the future of atomic energy research after the war was being considered. In the late summer of 1945, a recommendation to set up a British Atomic Energy Research Establishment was approved by the Government and an aerodrome at Harwell, Berkshire, was chosen as its site. It was also decided that responsibility for the work should be transferred to the Ministry of Supply. The Ministry of Supply took over the project on 1 December 1945.

3.2.9 There is little doubt that the rapidity with which a scientific and technical task of this size was achieved will be regarded in the future as one of the greatest achievements in the history of scientific development, quite apart from the consequences of the achievement itself. Physicists, chemists, biologists, and engineers, of many nationalities, working in three countries, in the difficulties of wartime conditions and in the greatest secrecy, turned a scientific curiosity into the greatest material force for good or evil which the world has ever known. With the political and economic consequences of the release of atomic energy, we are here not concerned, but the history of this research demonstrates, most vividly, to everyone, both scientists and laymen, the way in which results of the greatest practical significance may emerge from academic investigations; how the development of such discoveries can be achieved by planned research involved the cooperation of workers in all branches of science and in several countries; and also how the results of scientific research, neutral in themselves, are capable of being developed for good or for ill.

4.3.3 What a Research Council Does

From the Annual Report of the Medical Research Council for 1960–61, Cmnd. 1783, July 1962, pp. 7–9.

Organisation for Research and Development
At the level of management the task of a Research Council is to reconcile two apparently conflicting requirements. On the one hand they have to try to ensure that the new knowledge required by society through Government is forthcoming and to this end have to distribute their support of research according to their assessment of present and future needs. On the other hand they have so to devise their conditions of support as to preserve to individual workers that intellectual initiative without which creative work cannot develop. The problem for a research organisation is thus to reconcile the implementation of a balanced and comprehensive research policy at the national level with the need for intellectual freedom at the level of the individual or the research team.

The solution of this dilemma is largely a matter of the scale of operation. If resources are restricted or one thinks only in terms of individual institutions or particular organisations, the dilemma is very real, and, in order to discharge its social obligations, a research organisation may then be forced to have recourse to control and direction, even at the risk of inhibiting in its workers the creative thought and initiative on which its efficiency ultimately depends. But if the scale of operation is adequate the difficulty need not arise; for if support has been spread widely and with foresight and if the central organisation commands the confidence of scientists in general, then the relevant spontaneous interest can practically always be found. In other words, the solution of the dilemma lies in drawing a distinction between direction of research policy and directing individuals. It is of the essence of policy to have an object; but in relation to creative work the path to that object lies not through prescription to individuals, but through the informed selection of projects to be supported.

The question of what is an adequate scale of operation for a central research organisation with comprehensive social obligations in its particular field is, therefore, of critical importance. Today the answer can no longer be in doubt. The demand for new knowledge is such that nothing less than the national scale will suffice, for it is only on this scale that the requisite variety of researches, interests and institutions can be found to allow the necessary freedom of initiative in policy.

The classification of scientific activities is not unrelated to this problem, for it is on our ideas on this matter that any subsequent observations or recommendations on detailed organisation will largely depend for their justification. If, for instance, it is accepted that reasonably distinct categories of scientific activity can be identified, then it might be feasible to provide for these in different agencies and for administration to define the policy for each. If, on the other hand, we are dealing with a continuous range of activities, then segregation is to be avoided and a correspondingly more fluid organisation must be devised.

In our view, there are only two qualitatively different types of scientific investigation: that which is directed to increasing our understanding of naturally-occurring phenomena – research – and that which aims at bringing knowledge so gained to the solution of human needs – applied research or development. Even these are little more than abstractions at either extreme of the range. Between them lie every shade and combination of activity, and it is this that makes organisation in science possible. No research worker is entirely divorced in interest from his fellows or indifferent either to the use that can be made of his work or to its theoretical implications. All range, to some extent, about the centre-point of their interests, provided that they are not prevented by segregation from habitual contact with workers whose interests are contiguous. It is a major task of organisation to promote such interplay. Applied research on blood transfusion would be hamstrung if separated from relevant investigations in genetics or immunopathology. Research in endocrinology would be grossly handicapped but for the experience available in clinical practice. Laboratory and practical work in nutrition mutually interact. The interdependence of chemistry, biochemistry, chemical pathology and the study of human metabolic diseases is evident. We are witnessing a regrouping of disciplines, a prominent feature of which is the disappearance of the previous barriers between the laboratory and practice. In our opinion, therefore, it would be undesirable, on either theoretical or practical grounds, to do anything that tended to separate research aimed at the applications of knowledge from the investigation of natural phenomena out of which this knowledge grows.

The preceding paragraphs have been concerned with the relation between a research organisation and the scientific world. Equally important, in modern society, is the relationship between such an organisation and the world of practical affairs, particularly that of Government. Reference has already been made to the far-sighted planning that has characterised organisation for science in this country. In this connection the Haldane report on The Machinery of Government (Cd. 9230, 1918) is outstanding.

This established as a principle that a central research organisation should be separate from the executive departments of Government; and this principle has more than justified itself in practice. Because of it the central research organisation has been accepted as a source of independent opinion in which both Government and public can have confidence. In 1955 Parliament referred to the Medical Research Council for an opinion on the hazards from nuclear explosions. After the Windscale accident Parliament and public demanded an independent inquiry into the effects on health, by the Council. During the period of food rationing the Council were accepted by the public and the medical profession as a court of appeal on its medical aspects. The Council's assessment of drugs and vaccines is accepted both by the public and by industry as independent. Investigations by the Council are frequently in demand by Trade Unions in regard to controversial questions on possible associations between industrial conditions and health.

It was to promote this confidence that the Council were made independent and given their present constitution. It is in conformity with this intention that the Council's membership takes no account of geographical or sectional interests and that the only representation entertained is that of scientific disciplines and personal merit. For a similar reason no employee is a member of the actual Council.

Lastly, we would comment on the sources of scientific information, advice and opinion open to us. The Council are advised by two Boards and a large number of special advisory committees. Service on these is voluntary, but their membership ranges over the whole gamut of investigators with relevant interests, including not only members of the Council's staff but many more drawn from academic and professional fields. Such wide contacts are essential if a central research organisation is to discharge its function. Again representation is by knowledge and scientific merit alone, and members who are in the Council's employ attend, not as such, but by virtue of these same qualifications.

4.3.4 The Need for New Research Councils

From the Report of the Committee of enquiry into the Organisation of Civil Science, under the chairmanship of Sir Burke Trend, Cmnd. 2171, October 1963, pp. 32–3.

[See also 3.12.]

65. It can be argued that, in the light of the close inter-relation between fundamental scientific research and technological development, the responsibility for promoting scientific research in the universities, particularly in the physical sciences, should be combined – as at present in DSIR – with the responsibility for promoting industrial research and development. After very careful consideration, however, we have reached the conclusion that this combination of two responsibilities in one organisation will become too heavy a charge. Though inter-connected, they are inherently different in nature and purpose; and this difference of function should, in our view, be reflected in a difference of organisation. We believe, therefore, that the promotion of scientific research in fields of study other than medicine and agriculture should become the primary responsibility of a body, or bodies, concerned solely with research of this nature and that these bodies should be constituted on the model of a Research Council rather than a Department. We consider that two bodies of this kind are required – a Science Research Council and a Natural Resources Council; and that their relation to the Minister for Science should be the same as that of other Research Councils under the proposals made later in this report.

Science Research Council
66. This Council would assume from DSIR the responsibility for supporting the relevant research in universities and similar institutions; and this would be its primary responsibility. It would, however, also take over responsibility for post-graduate awards in science and technology. But this does not mean that its activities

should relate only to 'pure' science. An increasing amount of research in applied science (including engineering) is carried out in universities; and it would be undesirable, even if it were possible, to differentiate completely between the pure and the applied. It is true that the body concerned with the promotion of industrial research which we propose in Part VI below will support, by grants or contracts, a substantial amount of university work (e.g. in engineering) which has clear industrial relevance; but it should not be regarded as the sole source of support for applied science, especially in the universities where much work of a long-term character is carried out. The Science Research Council should, therefore, include both pure and applied science in its purview; but on the applied side it should maintain a close link with the organisation for industrial research, just as it must also maintain close contact with the other Research Councils, in order to ensure that the field in each case is completely covered. No difficulties should arise which effective liaison between the agencies concerned cannot resolve. Indeed, there are certain advantages in an arrangement which would provide alternative sources of finance for research workers in new and unexplored areas; and, although we envisage that support for basic scientific research in the universities will in general be provided by the Science Research Council, this should not prevent the other agencies from continuing to promote or assist such work in their own fields.

4.4 RESEARCH FOR INDUSTRY

4.4.1 Factors of Industrial Development

From *Nature*, 12 November 1932, vol. 130, pp. 714–15.

Economic planning, in the sense that is now taken almost for granted, had its antecedents in the first world war. The Board of Trade set up a General Economic Department but this did not long survive the war. A second attempt was made in 1930 when the Economic Advisory Council was formed but the body seemed to be a victim of its internal divisions and it did not meet again after the summer of 1931. However, one of its actions was to form a Committee on New Industrial Development to examine the desirability of creating a new, central research organisation. The Committee reported in 1932 and came down against this proposal. The following extract is a comment upon the Report and it raises the whole question of the relevance of innovation in industrial change.

Without detracting from the praise deservedly extended to many of these Associations, it is possible to feel that the Committee on Industrial Development is a little unduly optimistic about both their achievements and future prospects. Few of the Associations are self-supporting, and in fewer still has there been any sign that the industry as a whole is prepared to shoulder the full burden of financial responsibility of its research needs. In consequence, there is uncertainty about the continuance of the Associations, and the initiation of large scale development work is a much less simple problem than that confronting the research departments of large units of progressive industry, such as the General Electric Co., Metropolitan-Vickers, and Imperial Chemical Industries.

The Committee is accordingly unconvinced that an important gap exists in our existing arrangements for industrial research, which would be filled by the creation of a new national research organisation to draw up programmes of research into the application in industry of ideas, inventions, or processes at present undeveloped and likely to remain undeveloped, and to institute the necessary researches. While admitting that notably in the older industries, much research which might usefully be pursued is at present left undone, the Committee suggests that the need would be met by the Department of Scientific and Industrial Research establishing a branch for the purpose of initiating the stimulation of research in such industries, and directs attention to the readiness of the Department to create organisations of its own in important industrial fields where it recognises the existence of a need for research.

The proposal for the creation of a new central research organisation rests on a distinction drawn by its supporters between research directed to the improvement of existing industries and that directed to the creation of new industries, the latter being normally responsible for increasing employment. While it is undoubtedly true that the application of scientific discoveries resulting in the creation of new industries has given employment to millions, under modern conditions the displacement effects are becoming more serious, so that the rise of a new industry frequently means the contraction of an older established one. So far as employment is concerned, the national importance of scientific research as a factor in determining the rise of new industries lies rather in the consequent ability of the nation to minister to the new needs, thus giving employment to some or all of those inevitably displaced by the disappearance of needs to which the older industries ministered. In the minds of certain leaders of industry, such as Sir Harry McGowan, to judge from his presidential address last year to the Society of Chemical Industry, there are even doubts whether violent fluctuations caused in industry by, among other factors, the sudden exploitation of a scientific discovery, are really beneficial; and in the type of planning towards which the world is tending, provision for fundamental research and the deliberate exploitation of its results either by the State or by industry seems essential.

Apart from this factor, it is at least open to doubt whether the distinction drawn between the two types of research is valid and whether scientific research can be directed specifically to the creation of new industries. We are still without a real technique of discovery, and important industrial developments as often as not are based on purely fortuitous discoveries. In general, the fundamental discoveries have resulted from patient and disinterested investigations, the quest of truth for its own sake. Where scientific research has produced striking results from its direct application to industrial purposes we have entirely different conditions: a mass attack by a team of specialists on a clearly defined objective in a field where a considerable body of accumulated knowledge already exists.

4.4.2 Planning Industrial Research

From J. D. Bernal, *Progress*, September–October 1934, p. 366.

Radical thinking in the interwar years frequently advocated a more coherent plan for the country's government research effort, and the progress made in the USSR was often held up as a model upon which the United Kingdom should base itself. This is an example of such advocacy.

Overlapping and gaps in the Department's scheme of things are not surprising, but no logical plan is practicable under present conditions. The object of the Department is not to build up a new and sound industrial structure by applying scientific method; its object is the modest one of stepping in here and there to remedy defects that arise in the normal course of working existing industries. By such small uncoordinated services the Department attempts to win recognition, and ultimately, endowment, from industrial firms. This restricted policy is not one to which scientific workers can readily subscribe.

The problems of industrial production call for the setting up of about a dozen central industrial institutes which would cover a large section of industry and act as the general clearing house for allied problems that occur in the different industries; such, for instance, would be institutes for metals, silicates, fibres, electricity, transport, etc. These institutes would be connected to each other through the Department of Scientific and Industrial Research, but each in its own way would act as a link between the more particularly industrial problems of the research associations, such as laundry, refractories, etc., and the more abstract and fundamental researches carried out in the Universities. The facilitation at the same time of the application of scientific discovery and of the development of new discoveries under the impetus of the problems of industrial practice, calls for a two-fold flow. In one direction we start with a fundamental discovery which has to be made concrete in some central technical institute and applied on a semi-industrial, then on an industrial scale; in the other direction we need arrangements by which the day-to-day problems of industry can be referred back to their more general origins, and ultimately stimulate advances of a purely academic nature. Such an organisation of research is no paper scheme – it is the basis of organisation of scientific research in the USSR where it has shown its value in spite of terrific technical and economic difficulties.

It is not enough to remedy defects in the working of industry as they occur, the new materials and new techniques must be brought together in as short a time of their discovery as possible and embodied into practice. Here, of course, the full difficulty of government research is apparent. If the government was to take a positive attitude towards research it would virtually be going into production on its own account, and yet if it is not done by the government it is either not done at all or there is an enormous time-lag between discovery and application. The lack of any easy channels of up and down communication between theory and practice does, in itself, produce a very natural feeling among industrialists against the pushing of fundamental research. This lack of connection shows itself also in the absence of an economic basis for applied research; no industrial process is understandable without knowledge of the conditions of supply of materials and the utilisation of products.

4.4.3 The Economic Value of Scientific Research

From Memorandum on the Finance of Science, presented to the Lord President of the Council by the Parliamentary Science Committee, April 1937, published in *The Scientific Worker*, November–December 1938, vol. X, no. 4, pp. 114–16 and 142.

The initiative for the presentation of this memorandum came originally from members of the research associations which, in the early 1930s, were faced with the possibility that the 'Million fund' [4.4.7] would soon be exhausted. The British Science Guild made the suggestion that a part of the tariff revenues should be devoted to the research councils. The Association of Scientific Workers subsequently took up the idea and, with the compliance of the Parliamentary Science Committee (see chapter 7), were able in 1937 to submit the memorandum to the Lord President. The paper was substantially the work of Professor J. D. Bernal. On the advice of the Advisory Council of the DSIR, which in effect said that things were best left as they were since they were unlikely to be changed for the better, the Lord President rejected the scheme. The first extract argues the tangible economic benefits to be derived from research. The second extract comprises the recommendations which accompanied the memorandum.

There can be no question as to the value of scientific research applied to industrial production methods. The whole of our modern industry bears witness to the effectiveness of that application. In particular, the industries of electricity, radio and aviation depend for their very existence on scientific research. Until the present century such research was largely carried out by individual efforts and without any subsidies from official bodies. In 1917 the Department of Scientific and Industrial Research was set up as 'a permanent organisation for the promotion of scientific research'. The Department, besides controlling the National Physical Laboratory, the Fuel Research Station and other central institutes, has created and supervised twenty joint Research Associations in different industries. The work already done in these has shown in the most concrete way how valuable organised scientific research can be for all modern industry.

Research carried out by the Department, or by the Associations, costing a few thousands of pounds, has, in several cases, saved the industries concerned hundreds of thousands of pounds per annum. This occurs as well in cases where some important technical process is discovered as in the more ordinary case where only a number of small improvements are made. Thus, the British Refractories Research Association by an improvement in the design of saggars, costing £8,600 in all, effected an annual saving to the pottery industry of £200,000 per annum, while a number of small improvements introduced by the Cotton Research Association, which has a total income of some £50,000 a year, amount annually to some £300,000.

Imperfect as this record is, it shows conclusively the enormous economic returns derived from the direct application of science to industry. From six research councils alone at the expense of less than £400,000 *in all* have come researches which have made possible in practice a saving of £3,200,000 per annum. The aggregate saving from the work of all the Research Associations must be several times this sum and must represent a return on money invested of between 500 and 2,000 per cent. And this is only part of the service to industry. The indirect effects are at least equally important both in regard to improving industrial practice and introducing new industries. Thus the metrological work of the National Physical Laboratory forms the basis of all accurate standardisations of engineering parts and the work of the Fuel Research Board is opening up new methods of utilisation of coal. We may indeed take as a measure of the value of scientific research to industry at least a large part of the enormous increase of productivity which we have seen in the past few years, as for example the fact that while in 1910 a ton of coal produced 550 units of electricity in 1934 it produced 1410 units, or more than double. There can be no question that if as much as one-tenth of the additional value created by scientific research could be appropriated for the use of further research it could be worked advantageously on a scale beyond the wildest dreams of a scientist of today.

It is more difficult to assess in concrete terms of money the value of Agricultural or Medical Research. The value of the work done is, however, beyond question. In improving the breeds of animals and plants, in combating their diseases and in devising new agricultural methods the Stations assisted by the Agricultural Research Council are doing essential and valuable work in safeguarding and improving the nation's food supplies and in helping the farmer over many difficulties. What Medical Research has done in helping to cure, and even more in preventing, disease is past measuring. As a single instance it may be mentioned that the first practical isolation of a pure vitamin was carried out at the National Institute for Medical Research.

In spite of this demonstrated financial benefit resulting from scientific research, the problem of its finance on a scale adequate to its maintenance and development is still an acute one. The reason is not far to seek. Scientific research does not pay in the same sense that a commercial enterprise does by producing goods or services of immediate value to purchasers. Carried on in an organised way for a whole industry or a group of industries, scientific research is effective not so much through the assistance it gives to this or that firm as by its value to the whole industry. There may consequently be little direct inducement to individual firms to subscribe to the cost of scientific research other than consideration for the general welfare of the industry or of British industry at large. There will indeed be a tendency to consider, even among those firms which recognise the value of scientific research by contributing to its development, that they are giving advantages to other firms and even to other industries which do not do so.

Expenses of this sort, though ethically desirable, cannot be economically justified in industries run on competitive lines. It is clear that the real value of scientific research, pure or applied, is spread widely and can be financed only by finding an equally comprehensive basis for the collection of necessary funds. The problem is essentially not one of soliciting money for some non-economic end, but of finding the means of collecting the necessary funds for scientific research from those who are likely to benefit from it and in some proportion to the benefits they are likely to derive.

Recommendations

A. 1. That the contributions from Government sources to the Research Institutions (National Physical Laboratory, etc.) and to Research Associations take the form of block grants for a period of five to ten years in advance.

2. That the Department of Scientific and Industrial Research, together with the Medical and Agricultural Research Councils, endeavour to obtain by negotiations with industrial firms, associations of such firms or other bodies, an agreement guaranteeing contributions to Research Associations, Research Stations, etc., for a corresponding period.

3. That the Department should open negotiations with industries for which research facilities are at present inadequate with a view to providing a comprehensive system of government-aided industrial research.

4. That the aggregate sum made available for existing and new scientific research, laboratories and investigations be graduated to increase annually according to a prepared scheme with allowance for extraordinary expenses.

B. 5. That in order to secure continuity and adequate expansion of scientific research (according to recommendations 1, 2, 3, 4), in periods of variable industrial prosperity, a *National Scientific Research Endowment Fund* be established.

6. That the Fund receive from the Exchequer an annual sum of £3,000,000 (or £2,000,000 – £4,000,000), or 10 per cent of the value of Customs Receipts.

7. That part of the annual receipts be used to defray the Government contribution to scientific research.

8. That the accumulated reserve of the Fund be invested in trust securities (except as under Recommendation 10) and the income thereof be used to defray part, and ultimately all, of the expenditure on Scientific Research.

9. That the Department endeavour to secure contributions from industry and agriculture to the fund which need not be regular in amount, but should provide over a number of years an aggregate equivalent to that provided from Government funds.

10. That legislation be amended to permit contributions or legacies from individuals or corporations to the

Fund to be free from Income Tax, Super Tax, or Death Duties, and that the Fund be permitted to accept stocks or shares in individual concerns.

11. That the control of the Fund and the payments from it to scientific research be vested in a Scientific Research Endowment Board, an autonomous authority with representation from Government Departments, Industries, Agriculture, Scientific and Medical Bodies, Universities, and the public.

NOTE: – Recommendations A (1, 2, 3, 4) are independent of the adoption of the scheme contained in recommendations B (5–11). If the latter scheme were adopted Recommendations A would refer merely to the control of expenditure for the maintenance and development of scientific research, and not for the provision of finance.

4.4.4 An Industrial Research and Development Authority

From the Report of the Committee of enquiry into the Organisation of Civil Science, under the chairmanship of Sir Burke Trend, Cmnd. 2171, October 1963, pp. 38–9.

[See also 3.12 and 4.3.4.]

87. In the preceding paragraphs we have outlined a more rational and comprehensive organisation for the promotion of scientific research, by universities and Research Councils, in the 'pure' or 'basic' bands of the spectrum. We turn now to its more 'applied' bands, i.e. to the promotion of research and development in relation to industry; and in the following paragraphs we examine the type of agency which would be most appropriate for this purpose. Questions of policy fall outside our terms of reference; but we have felt both obliged and entitled to assume that it will be the Government's purpose to promote industrial research and development more intensively than hitherto, as an essential element in the sustained growth of the economy.

88. At present, DSIR is the principal agency of the Government concerned with the promotion of the application of science to industrial processes. We have already indicated, however, the reasons for our view that the function of supporting university

research should now be separated, organisationally, from the function of promoting industrial research. These arguments are reinforced by our assessment of the needs of industrial research. We believe that considerable development of existing activities will be called for; that, in particular, the separate but related functions of DSIR and NRDC in the promotion of development contracts should be more closely coordinated; and that this work together with the other industrial functions of DSIR, should be conducted in close consultation with other Government bodies working in this field – e.g. the Atomic Energy Authority, the Ministry of Aviation, the Post Office – and with the National Economic Development Council (NEDC) and Departments concerned with industrial and economic policy. We think that the task involved is one of major and growing importance and that a new authority should be constituted to undertake it.

89. We have considered whether these responsibilities should attach to a Research Council or to a Government Department. We feel that, if research and development are to make an effective contribution to industrial growth, the responsible agency must be more executive in nature than a Research Council. At the same time we believe that the techniques required in this field are more likely to be operated freely and without inhibition by an independent organisation than by a Government Department. We believe also that the success of any agency responsible for promoting the industrial research will depend, to a considerable extent, on its being able not only to recruit staff who have both scientific expertise and wide experience of industrial management but also to organise a good deal of systematic movement between its own staff and industry. The flexibility required in terms of both staff recruitment and staff movement is more likely to be achieved by an independent body than by a Government Department. We have therefore reached the conclusion that a new autonomous agency should be created to promote industrial research and development and that it might be called the Industrial Research and Development Authority (IRDA). This new Authority should stand in the same relation to the Minister for Science as the Research Councils. Its essential function would be to discharge, in partnership with industry and more intensively than at present, the functions of both DSIR in its industrial aspect and the National Research and [*sic*] Development Corporation (NRDC). It will be of great importance

that universities and Colleges of Advanced Technology should be closely concerned with the work of the Authority.

4.4.5 A British Research and Development Corporation

From Ministry of Technology, *Industrial Research and Development in Government Laboratories: a New Organisation for the Seventies*, January 1970, pp. 10–15.

This 'green paper', that is to say one intended as a basis for discussion and not a firm declaration of policy, appeared shortly before the general election of that year. It marked the culmination of the Labour government's thinking on the research establishments and a definite move away from earlier, centralising tendencies. Much discussion on it, largely critical in character, occurred in the spring of 1970 but with the change of government the BRDC proposal was effectively killed.

Current Problems

22. The simultaneous development of work for industry which is taking place within the Ministry of Technology's own organisation and in an independent statutory corporation (the AEA) raises the problem of how to avoid overlapping programmes and conflicts of interest. Although these difficulties are being tackled and may be overcome, it is difficult to envisage a completely satisfactory solution unless the resources can be brought together under the same management regime: only then could programmes be fully rationalised and the greatest benefits be secured. Rationalisation therefore requires a reshaping of the Government's industrial research and development resources by bringing them together in one organisation.

23. To set up such an organisation inside the Civil Service would be a move away from industry and would run counter to the 'hiving-off' approach the Fulton Committee favoured. A separate body, outside the Civil Service would be better able to rationalise programmes, and to operate more flexibly, and would be freer to associate its research and development resources more closely with industry.

25. A new organisation would have to tackle a more fundamental defect in present arrangements. Experience in this country and

abroad suggests that no Government department can decide centrally what research programmes are best designed to serve the needs of industry. As a general rule, only the 'customer' knows what he wants, and by his readiness to pay for it makes the 'supplier' aware of his requirements. This proposition has to be qualified for the reasons given in paragraphs 9, 10 and 14 above, but it is of crucial importance. A contractual relationship between Government laboratories and their 'customers' is essential if the programmes and size of these establishments are to be directly related to real needs.

The proposed solution

26. For all these reasons – the need to rationalise programmes in AEA and Ministry of Technology establishments, the need to link Government laboratories more closely with industry so they can better understand and more readily help to solve some of its problems, and the need for work to be undertaken increasingly on a contractual and commercial basis – the Government is considering setting up a new organisation, outside the Civil Service, in which all these research and development resources would be brought together under a single management.

27. The proposed new body would be a statutory corporation, possibly called the British Research and Development Corporation (BRDC). Its aims and functions would be:

(i) to encourage and support the development and application of innovation and technological improvement in industry for the benefit of the UK economy; and to carry out research and development for this purpose, both itself and in collaboration with industry and on repayment;

(ii) to carry out research programmes necessary in the public interest, including basic research, and other specific programmes of work required by Government departments and other public authorities;

(iii) to exploit where appropriate innovations resulting from Government-financed programmes carried out by other agencies.

28. BRDC would comprise the existing resources of the AEA's Reactor and Research Groups, of the Ministry's main Industrial Research Establishments and of the National Research Development Corporation. It would be a powerful organisation, at the

outset employing between 4000 and 5000 professional staff with a gross expenditure of about £70M annually. The numbers of professional staff and budgets of the separate organisations are given in the Annex.

29. Government work, including Government work of an industrial character such as the Reactor R & D programme, would inevitably comprise the bulk of BRDC's programme and would provide the greater part of its revenue for some years to come. Basic research, advisory services and statutory work might be funded by a Grant-in-Aid, but specific tasks or projects for Government departments would be charged at full costs. A gradual change to a contractual relationship should help to bring about a marked change of attitude within departments, which will have to pay for the work they want done, as well as within the Corporation, which will have to satisfy its customers. Financial expenditure would be reallocated between the Ministry of Technology and the departments which would assume responsibility for the cost of R & D projects, but there would be thereby no increase in total Government expenditure.

31. The Minister of Technology would appoint the Board of the Corporation and the Ministry would finance such basic research programmes as seemed desirable. Adequate arrangements would be necessary for the Department to determine or guide the broad lines of programmes, but the Corporation would be free to undertake on its own initiative work or investments on which it expected to recover its costs. It would however be required to operate within the general framework of the Government's industrial policies. Special arrangements would be made to ensure that regional considerations would be taken into account in BRDC's activities, especially in relation to the location of its establishments.

32. Given this degree of freedom, BRDC would be able to engage in joint ventures with industry, sharing costs and risks. The inclusion within this new Corporation of the NRDC would assist the development by the new body of an appropriate outlook to exploitation of new developments and bring about the benefits which have been discussed within paragraph 19(c) above. In effect the inclusion or integration of NRDC into the new Corporation would provide the essential element in its management structure which directed its attention to the need to secure a return on its

work and investment. There would, of course, be no restriction on BRDC's activities in exploiting inventions made by other bodies, as NRDC does now.

34. The Corporation's work for industry should clearly be conducted on a contractual basis, on repayment or as jointly funded ventures, because industrial participation is the simplest test of the value to industry of the work in Government establishments. It is not possible to estimate exactly how much of such work industry would want; but if a new research and development organisation were set up which would live to a growing extent on its earnings from contracts, joint ventures, royalty arrangements, etc., its success in meeting the needs of industry would play a part in determining the eventual level at which it operated. Such success would, over a period, make it possible to reduce the level of Government subsidy for industrial research, and the extent and nature of such work undertaken in Government laboratories would thus be increasingly determined by industry itself.

4.4.6 What about Research for Defence?

From 'Loose Edges in the Grand Design', *Nature*, 21 February 1970, vol. 225, pp. 673-4.

Many commentators remarked on the absence of the government's defence research establishments from the Ministry of Technology's Green Paper. This passage from a leading article touches on this.

The defence research establishments are a harder nut to crack. The Royal Aircraft Establishment and the Royal Radar Establishment are the two most obvious omissions from the proposed arrangements, but several other smaller laboratories should belong as well. For one thing, it is fair seriously to doubt the viability of the new organisation for research and development if these laboratories are not included. What will happen, for example, if under the new arrangements, the board of the British Research and Development Authority considers that it would be prudent to take on work in aerodynamics or telecommunications, thus threatening the defence laboratories with trespass? Will somebody say no? And if not, is there not a risk of serious muddle? But, worse still, without the skill and special knowledge of these defence laboratories, is there not a danger that the British Research and Development Corporation will be less well equipped than it should be to seize opportunities for which it is supposedly being created? Finally, it is also almost beyond dispute that the defence laboratories will suffer in the next few years the tribulations of finding that they have somehow outlived the purposes for which they were originally established. What will happen then? Will there have to be a second authority for research and development?

4.4.7 The Announcement of the Research Association Scheme

From Department of Scientific and Industrial Research, *The Government Scheme for Industrial Research*, June 1917.

The scheme was introduced near the end of the first world war in an attempt to overcome the situation which had existed at the outbreak of war when the technological superiority of German industry had been so glaringly demonstrated. The intention was that for each £1 contributed by industry the government would do the same up to a maximum of £1 million, after which government support would lapse and the associations would be enabled to 'take off'. The scheme was a mixed success and by no means all branches of industry set up research associations. Attempts to end the need for government aid never succeeded (the £1 million was finally all spent by 1932 at the worst possible time from the associations' point of view) and it has continued to be given until the present, although it is arguable that the associations have still to find for themselves their optimal role [cf. 1.3.8].

The Government, as already announced, have placed a fund of a million sterling at the disposal of the Research Department to enable it to encourage the industries to undertake research. Much will depend upon the way in which this money is spent. The independence and initiative of the British manufacturer have contributed largely in the past to his success. After the war he will need all possible assistance in undertaking and developing research work as a means of enlarging his output and improving its quality. But if the help is to be effective, it must increase his independence and initiative. It must avoid chaining him to the routine of Government administration however efficient. It must be so given as to enlist his active support. To the initiative of the manufacturer, the improvement of old and the discovery of new industrial processes in this and other countries have largely been due. It has been the co-operation of progressive industry with science which has led to the practical application of the results obtained in the laboratories of scientific men. The Advisory Council for Scientific and Industrial Research have therefore recommended after consultation with manufacturers and others that the new fund should be expended on a cooperative basis in the form of liberal contributions by the Department towards the income raised by voluntary associations of manufacturers established for the purpose of research. By this method the systematic development of research and the cooperation of science with industry will be carried out under the direct control of the industries themselves. It is also hoped that the cooperation of the firms concerned in any one industry may enable research work to be undertaken which could not have been dealt with by individual firms.

All considerations point to the necessity for combination to this end. If the firms in an industry which are engaged in the produc-tion of similar articles, or alternatively if the firms in different industries which make use of the same or similar raw or semi-manufactured materials will combine to improve those articles or materials, the Department will contribute liberally to a joint fund for this purpose. The fund for each industry will be expended by a Committee or Board appointed by the contributing firms in that industry, and the results obtained will be available for the benefit of the contributing firms.

The Department realises that there will be many difficulties in the way. Some industries are more homogeneous in character than others and could therefore agree more easily on the lines of investi-gation to be followed. Some industries are from their nature only interested in research because they are users of certain materials which are susceptible of improvement and these materials are also used by other industries. In some cases, accordingly, it may be possible to establish a Research Association for a whole industry, in others only a section may usefully combine, while in still other cases the most likely line of advance may be found in an association of firms in a number of industries which happen to use the same materials. It appears to be certain, however, from these considera-tions that some firms will wish to belong to more than one com-bination for research purposes, and that if wasteful effort is to be avoided every possible means must be used to encourage co-operation between different research organisations as well as between the several firms within the same organisation. . . .

The Government contribution will be promised for a period of years to be agreed upon, not exceeding five, though it may be extended if funds are available and the condition of the industry calls for further aid. The contribution will be made in the anticipation that when the new organisations are once fairly launched on their career the need for direct State assistance will disappear, and British Industry will be as self-sufficing in the field of industrial research as it has proved itself to be in other spheres of work.

4.4.8 Effectiveness of the Research Association Scheme

From the Organisation for Economic Cooperation and Development, *Industrial Research Associations in the United Kingdom.* Paris, 1967, pp. 124–6.

The success of the research association scheme taken as a whole is beyond question although it is difficult to measure. The research associations are all different from one another and, whatever yardstick be used to gauge success, it would be found that there is a very wide spread in the degree of success which could be ascribed to individual research associations. So uneven is the degree of success however that it causes concern. It is also difficult to generalise about the sort of industries in which most success has been achieved. As pointed out in Section 11, in successful research associations, the industries served may or may not be science based and may be big or small. Another feature which does not make generalisation easy is that research associations may be successful early in life and then seem to go through an unrewarding patch, possibly for many years.

The main value of the research association scheme, as it has turned out, is probably in industries with at least some research of their own and not, as originally thought, in industries with virtually no research of their own. Research associations serving industries composed of small units with no private research may be valuable but their contribution is usually different and somewhat limited. . . .

There has been criticism of the rate of application of the results of the research work of research associations. The task here is formidable and is particularly intractable where small industries or small manufacturers belonging to big industries are concerned. Part of the difficulty is the cost and risk of development although the fact cannot be emphasised too much that the application of much of the results of research does not require development work. The Special Assistance Scheme of the Department, which is now ending, was extremely successful in certain research associations by helping to close the gap between the circulation of research results and their application in industry, but cannot be considered as having done more than scratch the surface of the problem. A massive increase in facilities for explaining research results, whether these are from the research associations' own work or from other sources, is needed. There are various plans, some drawing on the staff of technical colleges, for making a new approach to this problem.

It has also been stated that research associations all too frequently develop processes or products which are not economic for industry to adopt. There is little evidence for this but naturally errors of judgement can be made and it is clear that research associations, being one stage removed from industry, have to be particularly careful in assessing the likelihood of their ideas being commercially viable.

4.4.9 Help for Inventors

From the National Research Development Corporation Annual Report for 1957–58, House of Commons Paper 34, December 1958, pp. 1–2.

The Corporation was established in 1948 under the Board of Trade by the first postwar Labour government with the express intention of granting financial aid for the development of inventions and for assisting in their exploitation in the public interest. Perhaps NRDC's most notable 'case' was the hovercraft. In this extract the Corporation outlines its philosophy.

3. In the earlier reports in the series of which this is the ninth we included a number of general comments in which we reviewed the task which had been entrusted to us and the nature of the problems we had to solve. When we had said all that could be said on the

subject we confined subsequent reports to purely factual statements of what had been done during the relevant years. This procedure culminated in our eighth report issued last year in which we summarised and indexed all the development projects in which we had been involved and gave brief descriptions of each. In this section of the present we revert to a general approach with particular reference to the general benefit which the community obtains through the Corporation's existence. We are concerned accordingly to summarise for the purpose of illustration a number of projects which have been successful. Our criterion for success is not necessarily that the Corporation has made a financial profit on the transaction, for this may not be determinable in cases where the Corporation's interest is represented by a revenue-earning investment based on some agreement stretching many years ahead. Our interim criterion is rather that there now exists some state of affairs, more or less taken for granted by that section of the community concerned with it, from which the nation as a whole derives an indisputable benefit, and which would either not be the case, or which would not have come about for a period of years ahead, unless the Corporation had taken the initiative under the statutory powers granted to it. In the simplest of these cases the Corporation has spent no money and received no revenue: it has merely played the part of a catalyst. In other cases it has made contingent grants by underwriting worthwhile projects which have found support from other quarters so that the guarantee has never been drawn on. In a third group the Corporation has for a modest expenditure held the fort while support was enlisted from other quarters. In a fourth group it has committed itself to a significant level of expenditure before the profitability or otherwise of its activities could be calculated, believing that a gain of a year or two in starting something would be of benefit to the community and that support from elsewhere would materialise in due course. Finally of course there are the majority of our projects where the initial prospects of a successful pay-off were good, varying from those which could reasonably be expected to pay for themselves but no more, to those which might yield a substantial surplus and pay not only for themselves but for other classes of projects in addition.

In undertaking these projects we have always had to scrutinise our terms of reference to see whether the project belonged properly to the Corporation or to some other body, a spending department for instance, or industry. Such issues are rarely clear cut. In doubtful cases we have felt it incumbent upon us to assume responsibility provided that some reasonable argument could be adduced for doing so. The fact that, given time, some still more reasonable argument might emerge for another body acting in our place, has seemed to us irrelevant.

5 Manpower Resources

The theme illustrated in this chapter is the responsibility assumed by the government since the end of the second world war for ensuring an adequate supply of trained scientists and technologists. Our interest is mainly with forecasting methods: attempts to match predicted demands with predicted flows, rather than with the nature, extent and quality of technical education. The following extracts have been chosen partly to bring out the growing sophistication in this period of forecasting techniques. The state of the 'numbers game' (which is what in retrospect it largely seems to have been) at any one time is not our object, but a prefatory note may help to place the extracts in context.

The total university population in the United Kingdom throughout the later interwar period was roughly constant, at about 50,000. This figure was about one-third higher than it had been shortly after the first world war. However, the enrolment of science and technology students had not increased in proportion during this period; at the end of the 1930s the total of almost 13,000 students marked an increase of only about 7 per cent on the enrolment in 1920. The output of qualified manpower just before the outbreak of the second world war has been reckoned at about 5,000 annually – made up roughly equally of scientists and technologists and taking into account all valid types of qualification. The table below shows only degrees awarded but it does serve to show something of the size of the postwar achievement.

New degrees* awarded

	1938–9	1947–8	1959–60	1963	1967
In science	2,167	3,972	6,570	7,559	11,264
In technology	1,048	2,306	3,415	3,466	6,563

* Other qualifications are omitted in order to ensure a measure of comparability.

The first extract below [5.1] is an indication of the state of the national awareness of its manpower resources on the eve of war. One concrete result of the war was the compilation, on the recommendation of the Royal Society, of a Central (Technical and Scientific) Register at the Ministry of Labour;* by the end of the war this showed a figure of 45,000 scientists.

The first major postwar inquiry into the state of the country's scientific and technical manpower was that chaired by Sir Alan Barlow in 1945–6 [5.2]. The introduction to this report (para. 2) could stand equally well as a justification for all the succeeding enterprises in this muddy field:

We do not think that it is necessary to preface our report by stating at length the case for developing our scientific resources. Never before has the importance of science been more widely recognised, or so many hopes of future progress and welfare founded upon the scientist. By way of introduction, therefore, we confine ourselves to pointing out that least of all nations can Great Britain afford to neglect whatever benefits the scientists can confer upon her. If we are to maintain our position in the world and restore and improve our standard of living, we have no alternative but to strive for that scientific achievement without which our trade will wither, our Colonial Empire will remain under-developed and our lives and freedom will be at the mercy of a potential aggressor.

Barlow forecasted a total demand of 90,000 scientists by 1955, but believed that the universities by then would be incapable of supplying more than a total of 64,000. In fact the predicted necessary doubling of science graduates was achieved by the beginning of the 1950s, much earlier than expected. However, the production of technically qualified graduates was below the requirement [5.4].

Before long it was recognised by the government that the Barlow target was far too low and in 1956 there was announced a further redoubling of the output of scientists and technologists, to be reached by the late 1960s [5.5, 5.6]. Shortages of technologists were again evident and thus, also in 1956, the government published a White Paper on technical education (Cmd. 9703) which proposed a new technological postgraduate degree and the promotion of ten technical colleges to be Colleges of Advanced Technology.

By the end of the 1950s the output of scientists and technologists, qualified by all routes, had reached an annual 15,000. This increase was reflected in an increasing proportion of full-time students in science and technology faculties: this rose from 35 per

* R. W. Clark, *The Rise of the Boffins* (Phoenix House, London, 1962), pp. 56 ff.

cent in 1956–7 to 40 per cent by 1961–2. By 1961 the Committee on Scientific Manpower was estimating that output and demand would be in balance by the mid-1960s [5.7].

Yet as if to demonstrate the pitfalls in this type of forecasting, in the following year (1962), the same committee produced another forecast indicating a near stationary output of engineers and technologists – to be set against an increasing demand [5.8]; furthermore the surplus prophesied in 1961 had turned, a year later, into a large expected shortfall by the middle of the decade.

From the late 1950s on there had been a growing awareness that the country was steadily losing qualified scientists and engineers, mainly to the USA; this flow was gaining in numbers as the following figures show.

Emigration from the UK

	Scientists	Engineers, technologists
1961	1,300	1,920
1963	1,490	2,475
1965	1,810	3,255
1966	1,975	4,240

What was a cause of particular anxiety was that the men already in greatest need were the ones apparently most prone to leave. The Royal Society had already alerted public opinion when in 1963 it discovered that nearly one-eighth of the annual output of Ph.Ds was emigrating. In the years immediately following concern escalated rapidly as the general public resolutely refused to recognise the phenomenon as world-wide and also as having a long history [5.12]. Finally it was realised that the government could do little about it; the flow was ultimately damped down by modified immigration laws and changing economic circumstances in North America.

Throughout much of the postwar period it had been repeatedly argued that the supply of engineers and technologists was inadequate. Evidence presented by M. J. Peck* shows that this is no more than an overt indication of a situation going back at least to the beginning of the century. British universities have traditionally given the impression of favouring scientific as opposed to engineer-

* In *Britain's Economic Prospects* (Brookings Institution, Washington, 1968), pp. 448–84.

ing studies and thus industry* has had to compensate for the lack of graduate technologists by, as Peck puts it, 'the use of personnel in relatively more abundant supply (that is, substitution of scientists for engineers and less skilled labour for both)'† – in other words, technicians and craftsmen trained largely by apprenticeship.

The earlier manpower surveys [e.g. 5.2] did not adequately distinguish between scientists and engineers with the result that the extent of such substitution was effectively disguised,‡ but the Fielden report of 1963 [5.9] made clear the results of these types of substitution: poorer standards of engineering design with their consequential economic feedback.

As the 1960s drew to a close new issues and problems appeared: university science departments were showing more and more unfilled places. This was thought to reflect a disenchantment with science among the young, mirroring a widespread disaffection in university circles – a problem probably beyond the powers of any advisory body to solve. At the same time the qualified men and women that were coming forward were reluctant to go into industry [5.10, 5.13]. But there was some evidence that the job market was easing.

Finally, early in 1971, came some official confirmation through the use of improved statistical methods, that in one area at least, the brain drain, many of the earlier anxieties had been groundless.§

5.1 Training and Employment of University Graduates

From *Nature*, 8 January 1938, vol. 141, p. 51.

The editorial indicates the primitive state of manpower planning shortly before the second world war.

If happily unemployment of graduates is less serious in Great Britain than in some other European countries, it has been

* Insofar as it recognised the need for technically trained manpower. A survey by the Federation of British Industries of industrial research in 1959–60 showed that major branches of industry, but especially those dominated by small companies, employed very few research staff, e.g. food, instruments and textiles.
† *Op. cit.*, p. 451.
‡ A comparison with the situation in the USA is provided by the fact that in the mid-1960s the professional research force in the USA contained 76 per cent of engineers, but engineers accounted for only 57 per cent in Britain.
§ Department of Trade and Industry, *Persons with Qualifications in Engineering, Technology and Science 1959 to 1968* (HMSO, 1971).

sufficiently widespread to offer obstacles to the work of the Society for the Protection of Science and Learning in finding occupation for displaced men of science from abroad, apart from the existence of a by no means negligible amount of wastage and misplacement of young graduates of our own universities. Indeed, the position led the University Grants Committee in its last report to suggest that the universities themselves should consider not merely the question of the appropriate number of university students in relation to the population of the country as a whole, but also that of the appropriate number for any particular university. Granting that a university should not be too preoccupied with problems of purely professional vocations for its students, it should none the less face the question whether the time has arrived, or is approaching, when additions to its numbers will tend to impair the quality of the instruction and the value of the training for life it sought to give. . . .

It can scarcely be denied that neither the universities nor professional associations of scientific workers, despite serious warnings, have given anything like the attention to this question that it demands. The report on graduate employment recently issued by the National Union of Students, to which reference was made in discussions at the British Association meetings at Nottingham, brings together in this field a number of suggestions or proposals which require attention and action by professional associations as well as by university authorities.

This report, which brings out the fact that in Great Britain the question of misemployment among graduates is as important, if not more important, than that of actual unemployment, reveals the meagreness and inadequacy of existing statistics and the complete absence of planning or control over the numbers entering various branches of study, as well as the ignorance on the part of students proceeding to a university of conditions in the employment market. The most serious criticism in the report is, in fact, that of the lack of guidance and help available at the universities for placing students in careers. Even where appointment boards or officers are to be found, their organisation often leaves much to be desired. Contact with students is imperfect, and with industry, commerce and the professions is often poor.

5.2 The Longer Term Problem of Manpower Supply

From Report of a Committee Appointed by the Lord President of the Council, under the chairmanship of Sir Alan Barlow, *Scientific Manpower*, Cmd. 6824, 1946.

The first of the major postwar manpower reports.

(1) *Supply and demand*

10. To arrive at our existing capital of qualified scientists we have had made an actuarial calculation based on the ascertained output of qualified scientists over a number of years and the recorded death rate for these years. In such a calculation a fundamental factor is the average active life of a scientist, the length of which can only be estimated after allowing not only for the easily assessable incidence of retirement on grounds of age and of death but for other forms of leakage from the profession as well. In the Civil Service the average professional life is 30 years, and we have used 30 years in making our calculation. As a result we arrive at a figure of 60,000 as the absolute maximum potential capital of qualified scientists in 1946.

11. But there can be little doubt that the young Civil Servant who has already chosen his profession is more likely to remain a Civil Servant than the student who has just passed his final examination in science is to remain in definite scientific employment. Moreover, in our calculations some allowance must be made for graduates who came from overseas and returned home after securing their degrees and also for people who only studied science in order to pursue other subjects, notably medicine.

12. We know that there were 45,000 scientists registered on the Ministry of Labour's Central (Technical and Scientific) Register at the end of 1945, and our existing capital must therefore be somewhere between 45,000 and 60,000. The Ministry of Labour inform us that, to the best of their belief, the Register covers between 80 per cent and 85 per cent of working scientists, and, in all the circumstances, we have come to the conclusion that it is unlikely that the nation has at its disposal today a force of more than 55,000 qualified scientists.

13. The assessment of future demand is no less problematical. We have studied the available results of a number of recent inquiries, among them one carried out by the Industrial Research

Committee of the Federation of British Industries on the probable demand for research and development workers in industry in the post-war years. As a result we have arrived at a figure of approximately 70,000 as the estimated minimum demand for scientific workers in this country and in the colonial service in 1950. Of this total roughly 30,000 represents teachers in the Universities and secondary schools.

14. We are conscious that too much reliance cannot be placed on this figure and there is, indeed, reason to believe that it represents an underestimate (perhaps a serious under-estimate) of the number of scientists whom the nation could usefully employ once peacetime industry gets into full production and science begins to build upon the many advances that it has made under the stress of war. We hope, indeed, that this is so, as we hold strongly that there is even now insufficient appreciation of the potentialities of well-directed scientific effort. But for the reasons which we set out below we fear that it will prove to be beyond practical possibility to add even 15,000 qualified scientists to our existing capital within the next five years, and we have therefore thought it well that our estimate of future demand should be a cautious one.

15. It is only to the Universities that we can look for any substantial recruitment to the ranks of qualified scientists. The proportion that has come from other sources in the past is very small indeed and we do not favour any attempt to add a responsibility for producing a substantial number of pure scientists to the existing and prospective burdens of the Technical Colleges. Generally speaking, the university is an essential stage in a scientist's education and in any event the Technical Colleges will be hard put to it to produce the number of technologists that are required to support and apply the work of the scientists.

16. Before the war the British Universities were turning out on the average some 2,500 scientists each year. The rate of output fell in the late 30s but recovered during the war and the science faculties are now practically full. On the assumption that the faculties continued to be full but did not expand their output, they would turn out perhaps 12,500 scientists during the next five years.

17. The Universities have, however, already indicated that they are prepared to expand their output provided that the necessary finance is forthcoming. In May 1945 the University Grants Committee invited all Universities in Great Britain and the three

University Colleges of Exeter, Nottingham and Southampton to formulate an estimate of the expansion in their student body which, ignoring financial considerations, they would contemplate when they had returned to normal conditions. The estimates received by the University Grants Committee in reply to this inquiry show considerable divergencies among the Universities. Oxford and Cambridge, for instance, did not feel that they could expand their numbers at all above their pre-war level, while the English Civic Universities as a body thought that an expansion of the order of 85 per cent might be possible within the first post-war decade. Taken as a whole, the replies of the Universities envisaged a potential increase in student population during this decade of approximately 45 per cent over the pre-war strength. The figures given were not all sub-divided between the various faculties, but it is probable that, in view of the more elaborate equipment which it requires, science would not expand quite proportionately to the expansion in arts. Let us say that it would expand by about 40 per cent.

18. If we were to accept this estimate, we should come to the conclusion that, from a pre-war average annual output of 2,500 scientists, the Universities' estimates would yield by 1955 an output increased to 3,500 per annum.

19. The Universities were not asked to estimate the rate at which their expansion could be achieved, but for obvious reasons it is likely that, if things were left to run their course, the curve of expansion would rise very slowly in the first years of the decade. If, however, we can assume that the progression from 2,500 to 3,500 scientists per annum were made at a uniform rate and that in each year of the decade the number graduating would be 100 more than in the year before, the additional number of qualified scientists in 1950 would amount at most to 1,500.

20. Towards the 1950 requirement of 70,000 we have, therefore, on this basis a gross supply of 69,000 scientists. But so far no allowance has been made for wastage. Death and retirement will take its toll of our existing capital. We must assume that a proportion of trained scientists will go into other professions and we must also assume that a proportion of the annual output from the universities will be foreign students. We cannot pretend to have estimated the extent of this wastage accurately. Wastage during the war years was small but the circumstances were quite exceptional

and a return to a more normal rate must be expected. It is not unreasonable to assume that the trained man-power available in 1950 would be less than 60,000, perhaps no more than 55,000.

21. But a review of demand over five years is an inadequate basis for any recommendations regarding provision for the training of scientists where three years is the minimum period of university education, and it is therefore necessary to have regard to the movement of demand after 1950. In the absence of estimates, even of the provisional kind upon which our 1950 figures are based, we can only make the broadest assessment of the possibilities. We do know, however, that in education the effect of the raising of the school-leaving age by one year in 1947, the establishment of the County Colleges and the subsequent raising of the school age by a second year will result in a further steep increase in the demand for teachers with scientific qualifications. Between 1950 and 1955 an additional 15,000 teachers will be needed, the curve of demand flattening out after the latter year. In industry we estimate that the exceptional demands caused by the needs of reconversion will have been determined by 1950 and that thereafter any expansion in demand will reflect mainly the increasing application of science. A steady but not exceptional rise in the employment of scientists in central and local government service is probable. Taking all the known factors into account we feel justified in assuming that by 1955 the demand for scientists will be not less than 90,000.

22. Towards this total demand we can set our present capital and the new scientists to come from the Universities between 1946 and the end of 1955, less the wastage over those years. On the assumption that the Universities expand according to the figures that they returned to the University Grants Committee and, neglecting the effect of conscription, we make an approximate calculation of our new capital as follows:

(a) Present capital	...	...	...	55,000
(b) Additions to capital ...	...	...	30,000
				85,000

For wastage, in the absence of any better figure we may again assume that, in science as in the Civil Service, an average professional life lasts 30 years and make an allowance of, say, 3,000 for new science graduates who do not enter the scientific professions in this country. Hence:

(c) Reduction of present capital	...		18,000	
(d) Loss from new capital	...	...	3,000	
				21,000

leaving us with

(e) Net capital at 1955 ...	...	...	...	...	64,000

23. Thus against the 1955 estimated demand of 90,000 we could not expect more than 64,000 under present plans. In the face of these figures, we consider that, if national recovery and our material progress are not to be dangerously hampered by lack of trained scientific ability, the output of scientists must be raised very much above the level of the present University proposals. It is too early to attempt to estimate exactly what the output should be once the nation has settled down to peace-time conditions but we are satisfied that the immediate aim should be to double the present output, giving us roughly 5,000 newly qualified scientists per annum at the earliest possible moment.

(4) The Effect of our Proposals

(I) THE SUPPLY OF GRADUATES DURING THE NEXT TEN YEARS

57. It will be seen from paragraphs 20 and 23 above that, even on the most optimistic view of present plans, we are likely to be several thousand scientists short of our requirement in 1950 and 1955. In considering what effect the acceptance of our recommendation, that the output of scientific graduates should be doubled at the earliest possible opportunity, may have upon this shortage, it is necessary to take into account the following factors:—

(a) It takes at least three years to qualify for a university degree and a proportion of students must do a period of post-graduate work before they can be considered properly qualified. Even, therefore, if it were possible to double the intake of the Universities in October 1946 there would be no substantial rise in the output of qualified graduates until 1949 at the earliest.

(b) It is too much to hope that the Universities will in fact be able to double their intake in 1946. There will be more than enough applicants for places but we very much doubt whether, even assuming that teachers are available, the Universities will be able to carry through such a formidable

267

task of reorganisation in the few months available before the opening of the academic year. Moreover, while much can be done to solve the accommodation problems by recourse to temporary expedients, even the flimsiest building takes time to construct and the most suitable existing building, time to convert for university purposes.

(c) As far as recruitment to the Universities in the years succeeding 1946 is concerned we have set out above our reasons for thinking that this is the least serious of the problems which must be faced. We have, moreover, brought to the attention of the Government our view that, if the intellectual standard of graduates is not to suffer, it is desirable that science students should be allowed, if they wish and if there is room for them, to complete their full course and obtain their degrees before doing their period of military service. But military service must be done and, assuming that it lasts for a period of two years, it will mean that at any given time the equivalent of two years' output of graduates will be soldiers and not scientists.

58. If we take these factors into account, the only possible conclusion is that whatever is done to increase the output of science graduates, the nation will be seriously short of scientists in 1950 and that without heroic efforts it is unlikely that supply will have finally overtaken demand even five years later.

5.3 Manpower and Economic Growth

From the Annual Report of the Advisory Council on Scientific Policy 1951–52, Cmd. 8561, May 1952.

9. A policy for scientific man-power can in fact proceed on only one of two alternative assumptions. We can either assume that there will be a steady and rapid increase in our productivity, or that the economy will remain static and that productivity will increase only slowly. If we accept the first alternative, we must plan to meet the demand for scientists to which it will give rise, and on which it depends. If the second assumption is correct there is little hope of our remaining a great power, or even of our paying for the imports needed to sustain our economic life. If planning is to have any

purpose it must, therefore, be designed both to meet the needs and create the situation of the first hypothesis. The long-term trend is independent of any temporary over-production of scientists which at times might conceivably result from such a policy. Indeed, if an excess of supply is due to a recession in the economy as a whole, it ought rapidly to right itself, for the problems of a deflationary as of an inflationary crisis can best be solved by an increase in industrial efficiency, with a corresponding increase in the demand for scientists.

10. But in any event the current demand for scientists arising from industry, from the defence programme, from Government Service and from plans for assisting the undeveloped areas of the world is so great that we can see no prospect of supply exceeding demand for at least the next five years. There is no contradiction between our long-term aim and the problems of the immediate future.

5.4 The Need for Engineering Graduates

From the Committee on Scientific Manpower of the Advisory Council on Scientific Policy, *Report on the Recruitment of Scientists and Engineers by the Engineering Industry*, 1955.

29. Our inquiry has clearly established that a fear does exist in some quarters that the expansion of higher technological education and research by institutions of university status is detrimental to the interests of certain sections of the engineering industry. One of the main worries is that such men as go to the Universities to take up engineering are likely to stay there too long, and to study and to do research on topics of little value to industry. Another is that the increase in the size of the University population is 'creaming off' talent which would otherwise have entered industry directly as apprentices.

30. Widespread though these fears may be, the Committee on Scientific Manpower is not persuaded by the arguments from which they spring. It is first of all obvious that these views relate to only one, even if important, section of British industry as a whole, and even so that they do not represent the opinions of all the engineering firms concerned. Second, it is not only plain that engineering firms require good graduates as well as good apprentices, but that

however much the engineering departments of the universities may have expanded already, they are not, even on the basis of present demand, turning out enough men to provide our engineering firms with more than about three-quarters of the graduates they want. Thirdly, we are confident that, even if the universities were to expand further, and the number of students who wished to take engineering increased correspondingly, there would still be enough available talent in the population of boys and girls who do not go to a university to provide industry, over the next decade, with all the good apprentices which it might wish to recruit directly for its apprenticeship schemes.

31. We are very impressed by the fact that even a section of British industry, which has been concerned lest expansion of technological departments at the universities would deprive it of good apprentices, cannot recruit more than three in four of the university graduates it wants. Whatever happens to apprenticeship schemes, the demand by industry for scientists and technologists trained to graduate level is bound to increase at an ever-expanding rate in response to the rapid growth that is taking place in scientific knowledge. Some of the firms we approached appreciate this point fully; at the other end of the scale are those who neither yet want the university-trained scientist/engineer, nor indeed have much call for student apprentices. This will have to change if these firms, and British industry, are to maintain their present competitive positions. The future of our country is vitally dependent on advances in technological knowledge, and these must stem from research work at the universities as well as in industry and government laboratories. In any event, the scale of postgraduate research work in engineering which is contemplated for the expanded university departments as a whole is no greater than is necessary to ensure that these departments will thrive as scientific institutions.

32. We fully appreciate that engineers need a considerable amount of practical training in industry in addition to theoretical studies at the university. Student apprentices are given a combination of practical and theoretical training which can produce very good results, especially where the system is the 'Sandwich Scheme', with for example six months in each year fulltime at a technical college. At the same time it is clear that, to keep up to date with new scientific developments and with the increased complexity of industrial processes, there is an urgent need for the multiplication

of postgraduate instructional courses, as opposed to engineering research, at the universities. Here, as we have already emphasised, is where industry must get together with the universities to make its needs better known.

5.5 Method of Estimating Future Demand

From the Office of the Lord President of the Council and Ministry of Labour and National Service, *Scientific and Engineering Manpower in Great Britain*, 1956, pp. 13–14.

The Report sought to base its manpower projections upon the anticipated growth of the economy, considered by major industrial sectors.

43. Our estimate of the future requirements of manufacturing industry has been calculated, in the way outlined in paragraph 42, by first making an appropriate addition to cover prevailing shortages in the existing stock of scientists and engineers in each major industrial group. This figure was then increased in direct proportion to the cumulative rate of growth which each industrial group is postulated to undergo over the next ten years. The result gives an estimate of demand for scientific manpower in 1966, in terms of the number of scientists and engineers, which we believe would be necessary to permit an increase in total industrial output amounting to 4 per cent per annum.

44. The estimates of future employment of scientists and engineers in the public sector of scientific employment have been dealt with separately. The estimates for Government Departments are based on existing policies and allow for only a very modest increase over the next ten years. In the field of education we have postulated sufficient teachers to deal with the increased number in secondary schools over the next ten years, but not for any further concentration on scientific subjects nor for any improvement in existing staffing ratios. A reasonable allowance for the latter would have been about 4,000 additional Science Graduates. Requirements of the universities are assumed to increase by about 100 per cent over the ten-year period.

45. We fully realise that we may have taken too conservative a view of the demand for scientific manpower that may develop in some parts of the public sector of the economy. On the other hand, we

have not allowed for any decrease in our research effort for defence. On balance, it is reasonable to suppose that changes in demand by public authorities over the next ten years will not materially affect the order of magnitude of our final answer about the training of scientific manpower. Our concern is to provide an estimate of requirements which will be generally accepted as a minimum goal at which the universities and technical colleges can aim in the effort to provide for the increasing demand which is likely to develop in all sectors of the economy.

46. The broad conclusion of our calculations is that a condition essential to an annual rate of growth of 4 per cent in total industrial output is an increase in the number of qualified scientists and engineers employed from the present level of about 135,000 to somewhere in the region of 220,000 in 1966 – an increase of rather over 60 per cent. In the strict sense of the definitions given in paragraph 6, the increase in requirements of engineers is estimated at about 70 per cent between 1956 and 1966, while the demand for scientists is expected to rise by about 50 per cent over the same period.

47. The order of magnitude of this projected increase in requirements of scientists and engineers is influenced by several factors, of which the most important is the assumption that on an average the demand in each industry will increase in direct proportion to increases in output of that industry. We are, of course, aware that, in some industries, the number of trained scientific personnel is expected to increase at a faster rate than output – whether in an effort to catch up with present technical knowledge or, given that they are not technically backward, in adapting to as yet undefined technological developments. If this were to turn out to apply to British industry as a whole, our estimates of future demand would undoubtedly prove inadequate. But we have no reason to believe that this will be the case. Even though the same trend might well apply to certain backward sectors of industry, which in total are responsible for only a small part of the overall demand for scientific manpower, the most reasonable assumption, unless we turn entirely to guesswork, is that an average ratio of 1:1 strikes a balance between industries in which trained manpower is likely to increase at a rate faster than output and those in which it will increase at a slower rate. The method followed in our calculations allows for the weighting of the final answer according to the greater volume of

growth which it is assumed those industries (e.g. the chemical) that are now relatively strong in scientific manpower will undergo relative to others which are weak.

5.6 Correlation between Industrial Output and the Number of Scientists and Technologists

From Professor J. Jewkes, 'How Much Science?', *Advancement of Science*, September 1959, vol. XVI, pp. 72–3.

The extract is taken from the published version of the presidential address to Section F at the 1959 York Meeting of the British Association. Jewkes critically examines the forecasting assumptions made in the 1956 manpower report [5·5].

The only attempt known to me to establish a link between the number of scientists and engineers in employment and the growth of the economy and, in this way, to predict the future need for scientists and technologists is that of the Committee on Scientific Manpower.* Their argument is worthy of close scrutiny. There are three stages in it which can be set down in the actual words of the Committee's Report:

1. 'There is a definable relationship between the rate of increase of industrial production on the one hand, and the number of trained scientists and engineers employed by industry on the other.'

2. 'It is more reasonable to base our projections of the likely need for scientific manpower on the assumption that demand within each industry will, on the average, increase in direct proportion to increases in industrial output than on some even more arbitrarily chosen relationship.'

3. 'We were advised that it would be appropriate to work on the basis of an average increase of industrial production at the rate of 4 per cent per annum.'

If, therefore, the number of scientists and technologists is linked to industrial production in this way and if industrial production is to

---

* *Scientific and Engineering Manpower in Great Britain*, Office of the Lord President of the Council and the Ministry of Labour and National Service, 1956.

increase at the rate of 4 per cent per annum, then a basis exists for determining the need for the number of scientists and technologists in the future.

This approach is perplexing for more than one reason. Which is supposed to be cause and which effect in this correlation? Is it being suggested that the number of scientists and technologists is the crucial determining factor in economic growth, so that if we do not have a 4 per cent increase in the number of scientists and technologists we cannot possibly enjoy a 4 per cent increase in industrial production? Actually the annual percentage increase in industrial production between 1955 and 1958 was not 4 per cent; production did not increase at all in this period. But the number of scientists and technologists certainly increased. Does this mean that there were too many scientists and technologists produced in the period? If not, what becomes of the assumption that there is a definable relationship? If so, and if the crucial mistake made was in the assumption of a 4 per cent increase in industrial production per annum, what is the point of trying to predict the future annual requirements of scientists and technologists by basing it upon another prediction, that of the probable increase in industrial production, which is no easier to make than a straight prediction out of the blue of the number of scientists and technologists that will be required?

An even more puzzling point is that this Committee, along with other authorities, accepts by implication one assumption for which there appears to be little evidence and which is highly pessimistic in character. The assumption is that, since the number of scientists and technologists moves in step with industrial production, output per head of scientists and technologists will never improve, that whatever economies are made in the future they will not be in that particular type of labour. If this were true it would be a striking fact that scientists and technologists could economise in everything but themselves. I do not see why this assumption is made. And I suggest it is pessimistic because, if we accept the idea of a 4 per cent annual increase in industrial production, a short calculation shows that the number of scientists and engineers in Great Britain will have to increase from about 120,000 in 1955 to 5,171,000 in AD 2050. That is a lot of scientists and technologists.

5.7 A Surplus in Sight

From the Committee on Scientific Manpower and Statistics Committee of the Advisory Council on Scientific Policy, *The Long-Term Demand for Scientific Manpower*, Cmnd. 1490, October 1961.

The Committee adopted new and more refined forecasting methods and for the first time since the end of the war looked for the manpower supply being in surplus.

6. The problem of assessing demand has been accepted as essentially more hazardous than that of calculating supply. The method we used in 1956 (assuming in the case of industrial demand a direct ratio between growth in output and demand for manpower) has been superseded by a more detailed assessment of each sector of employment, and in the case of industry of the main industrial groups separately (paragraphs 22–67). We accept the reasons for this change of method. We are glad that the Statistics Committee has avoided restrictive assumptions, in view of previous tendencies to under-estimate both the capacity of the educational system for growth and that of industry to absorb manpower. Some sectors of industry may well use qualified manpower more extensively under conditions of freer supply. We are sure that it would be folly to take the risk of significantly under-estimating the demand for employment and of cramping the vital growth of our scientific effort by attempting to draw too fine a balance between supply and demand.

7. Demand for qualified manpower for all purposes has been included; where indications were available, the numbers have been estimated of scientists and technologists that employers will wish to engage irrespective of the kind of work on which they will be employed at various stages of their careers. The Committee has avoided an exclusively 'vocational' approach, and has taken account of the process, which may well increase in pace, whereby scientists and technologists will progressively invade fields of employment outside their disciplines, whether in management, the public service, or elsewhere. In this the tradition of previous surveys has been followed.

8. As regards industrial employment, the most important assumption made, on the basis of inquiries, is that the speed of absorption of qualified manpower in technologically advanced industries tends to slacken off when a density is reached at which

275

the technologies concerned are fully 'manned up' (paragraph 24). Thereafter the rate of absorption would tend to be governed by the size of the industry and by replacement until such time as far-reaching but unknown technological changes (which are apt to be long-term processes) present an entirely new industrial situation. The Committee had grounds for assuming that technological 'break-throughs' as such within a large industry would not normally involve those industries in taking on significant additional resources of manpower.

12. It turns out that, overall, the Committee's assumptions imply (Table 11) almost precisely the same rate of growth of scientific employment over the next decade as has taken place in the years 1956 to 1959. But the methods by which the estimates have been built up give a much greater sense of assurance than would a simple projection of an existing trend.

13. The conclusions, that by 1965 supply and demand of manpower should not be much out of balance and that a surplus may exist after that date, will come as a surprise when one considers the crippling shortages of scientific manpower that have been experienced since the war. It is recognised that the balance is a statistical one; and that supply and demand in individual disciplines (we would instance mathematics as perhaps the most important example) cannot so easily be equated. There may well be important differences in the factors which govern employment prospects between the specialised technologist and the engineer with a general training; and between engineers as a group and scientists. These problems remain to be examined. But it is possible now to foresee something of the fruits of the efforts which have led since the war to such a striking increase in our annual output of qualified scientific manpower; ten thousand in 1955; sixteen thousand five hundred this year; with a prospect of twenty thousand by 1965, and thirty thousand in 1973.

14. In our view the possibility that there will be a surplus of scientists over immediate 'demands for employment' should be welcomed. It should make possible a rational, as opposed to an emergency, use of the scientific disciplines. It should mean that at long last we shall have a supply of qualified manpower with a scientific training for management, administration, and the professions generally, in addition to those who up to the present have

been drawn inevitably into vocational employment. We do not doubt that scientific education will adjust itself to this new prospect; and that in the same way as only a proportion of those trained in the classics and history have expected to find employment in their own fields of study, an increasing proportion of those trained in specialised scientific disciplines will obtain employment outside them. We think that both science and the nation will benefit from this adjustment. We believe that there are already indications that Universities are thinking along these lines.

75. On the assumptions we have made, and subject to these reservations, we conclude that the overall supply and demand for qualified manpower will not be very much out of balance at the end of the first five years of the decade 1960/70. If anything, a slight shortage of technologists will be balanced by a slight surplus of scientists; any possible divergence of supply and demand comes into the second five years. We found no adequate basis for making separate estimates of the supply of or demand for particular kinds of scientists or technologists.

5.8 A Revised Estimate

From the Committee on Scientific Manpower of the Advisory Council on Scientific Policy, *Scientific and Technological Manpower in Great Britain in 1962*, Cmnd. 2146, October 1963, pp. 21–2.

The revised manpower estimates were greatly at variance with those produced only a year or so earlier [see 5.7].

46. In our 1961 Report* we were able to indicate for the first time that, although a shortage of scientific manpower was likely to continue for a long time in certain disciplines, the total supply of qualified manpower was beginning to approach the total of identifiable demand – in the assessment of which we made such allowances as the available information suggested were reasonable for demand in sectors of the economy not covered by the Manpower Surveys. These allowances have proved to have been too low, and the information now available to us from the 1961 Census inquiry, together with the results of the 1962 Manpower Survey and

* *The Long-Term Demand for Scientific Manpower*, Cmnd. 1490, October 1961.

employers' estimates of requirements, suggests that we are still some way from a satisfactory balance of supply and demand, particularly as regards technologists.

47. The relation between supply and demand differs greatly from one discipline to another. This has always been so, but in the conditions of extreme overall shortage which prevailed after the war, the first objective was to get near to an adequate total supply of scientific manpower. As we approach that first objective, analysis by discipline becomes increasingly important.

48. Moreover, as the 1961 census has shown, the range of employment of qualified scientists and technologists is considerably wider than the fields covered by our successive Scientific Manpower Surveys. Leaving aside postgraduate research students, the unemployed, and those employed in small industrial establishments, there were in April 1961, nearly 40,000 scientists and technologists working in fields outside the areas covered by the Survey. It is satisfactory that a proportion of these 40,000 qualified scientists and technologists must be engaged in the kind of occupation – management, administration and the professions generally – which was in our minds when we referred in our report of 1961 to the fields which would benefit from 'a rational use of the scientific disciplines'. We hope that the increased availability of qualified men will make possible a further substantial contribution of scientific manpower to these fields of employment, as well as to those with which we have hitherto been more directly concerned.

50. The prospects of employment for trained scientists and technologists are, therefore, good, and many will open in fields different from those which have traditionally been considered as the scientist's natural goal. Where vocational employment is concerned, more attention must be given to the question of balance, in both numbers and quality, between the scientific and technological sides. Substantially fewer are qualifying in technology as a whole than we expected in 1961, and serious shortages may, therefore, continue. This is to be explained mainly by the preference undergraduates have for the basic as opposed to the engineering sciences. There is also evidence, recently published elsewhere, that the average ability, judged in terms of 'A' level results, of those who take up engineering is lower than that of those who move to the basic sciences. This kind of intellectual discrimination against

engineering is notably absent in European countries. On the more favourable side, we have been pleased to note that more and more technologists are qualifying through university degrees, or by Diplomas in Technology acquired as the outcome of sandwich courses involving a succession of periods of fulltime study and related industrial experience, each lasting six months or longer. This reflects a growing appreciation on the part of industry that the attainment of professional qualifications by participation in part-time courses does not meet modern needs, and that periods of full-time study are essential if prospective technologists are to have the best possible grounding on which to develop their abilities in later life.

5.9 A Change of Emphasis in Engineering Education

From the Report of the Committee on Engineering Design, under the chairmanship of G. B. R. Fielden, FRS, Department of Scientific and Industrial Research, 1963, p. 28.

The Fielden Committee were concerned to demonstrate the importance of design in the engineering industries, but inevitably they were compelled to comment too on the quality of engineering education available in the United Kingdom.

85. Our main criticism of engineering education in the United Kingdom does not relate to the length of courses, nor, particularly, to their content, but rather to the emphasis given to research rather than design as the objective. It is probably right that the ambitious young scientist should be directed towards research and that his teachers should engage in this activity when not teaching. In consequence the rating of science teachers and their promotion prospects tend to depend upon their achievements in research. This bias of scientific education has spread to the university engineering departments and from them to the colleges of advanced technology. Teachers of engineering subjects in universities and colleges engage in research or aspire to do so and students naturally emulate their mentors. Only rarely, in universities, are students encouraged to regard their studies as preparatory to a career in industry as designers and producers of goods.

86. It is therefore not so much a question of 'teaching design' in universities and colleges, as of inculcating a different outlook among teachers and students on the uses to which their education

may be put. In particular, if students are to become designers, they must be given, and know that they are being given, the intellectual tools for a creative activity, and they must somehow be encouraged to acquire a taste for this activity.

87. The function of a course at university level should be to give the engineer an adequate grounding in the basic scientific principles applied in engineering, to teach him scientific method, to develop his critical faculties and to make him aware of the kind of problems he will face as an engineer. Experience in the United States of America has shown that the attempt to treat engineering as a branch of science in university courses has not been successful in producing engineers for industry. Scientific problems are assumed to have single solutions; the solutions to engineering problems are almost always compromises. While the engineering student must learn the methods of scientific analysis, he must also learn something of economics and of production methods which enter into every practical engineering problem, and develop a faculty for improvisation and invention.

5.10 Is Industry Being Starved?

From the Committee on Manpower Resources for Science and Technology, Interim Report of the Working Group on Manpower Parameters for Scientific Growth, Cmnd. 3102, October 1966.

By the mid-1960s a new worry had appeared: was the trained manpower being produced going into the most productive sectors of the economy? The evidence suggested that industry and, to a lesser extent, the school teaching profession were being denied their demands by the greater attractions of universities and government service.

Conclusions

33. The evidence we have assembled indicates a number of general trends, on which we comment below. We must however emphasise a number of points at the outset. Firstly, our conclusions are necessarily provisional, and we are well aware of the need for further statistical work. Secondly, it is evident that the position varies markedly as between science on the one hand, and engineering and technology on the other. Even within these sectors there are variations. The pattern of employment in chemistry is very different

from that in physics. In civil engineering there is, in all probability, still too little research, and the majority of graduates go straight into industry. Finally, the position varies strikingly with the quality of graduates. The trend away from industry is most marked in those with the highest academic attainments, and least apparent amongst those with the lower class of honours degrees or with ordinary degrees. Nevertheless, such trends amongst the best graduates inevitably suggest the pattern of first choice and ambition amongst the majority.

34. The pattern of flow in recent years is of a high proportion (64 per cent of those obtaining first class degrees between 1962 and 1965) of the most able graduates (as judged by academic standards) taking courses leading to a higher degree. Thereafter, with the exception of chemistry and some of the schools of engineering we have examined, the majority have either remained in the higher educational system or have entered Government research establishments; and a high proportion have emigrated to the United States. At the same time there has been a fall in the proportion (and in some disciplines in numbers also) entering 'Industry' both at the first and higher degree level, at the higher degree level from 529 (28 per cent) in 1960 to 510 (19 per cent) in 1965. Intake to 'Schools, Colleges and Teacher Training' has fluctuated about 17 per cent at first degree level and about 6 per cent at higher degree level.

35. This is in contrast to the pattern of demand as revealed by the 1965 Triennial Manpower Survey, even when allowance is made for new supply through other routes such as the Professional Institutions. The stated net additional requirements in 1968 are:

> Industry: 30,000.
> Schools and Colleges: 8,000.
> Higher Education and Research: 5,000.
> Government: 4,000.

At present rates of growth employment in the first two sectors would fall short of stated demand, and in the last two together could exceed their estimates.

36. The pattern of demand yielded by the 1965 Survey is not new, being broadly the same over the past decade. The inability of the schools and of industry to meet their stated demands are persistent features of this period, in contrast with the 'Higher

Education and Research' and 'Government' sectors which have on occasions achieved or exceeded their stated demands. While the total number of graduates in engineering, technology and science falls below the total need, as at present, a problem will remain. The requirements of Industry and Schools set out in paragraph 20 cannot be met by redeployment of the expected number of graduates without serious consequences for the 'Higher Education and Research' sector. There is a clear need to encourage more boys and girls to prepare themselves for science or technology courses in higher education and we hope that the inquiries initiated as a result of the Interim Report of Dr Dainton's Inquiry will yield definite suggestions on how to achieve this.

37. Since, however, it must be a considerable time before the total number of graduates is adequate, attention must be given to the imbalance in the present pattern of employment. The shortage of candidates for the total number of posts likely to be available has encouraged the emergence of a general attitude among students of regarding the 'Higher Education and Research' sector as very markedly preferable to the others, so that the student tends to enter the university already convinced that pure research is the goal worth setting his heart on. This attitude is of course not new, but if allowed to persist could have serious consequences. With too few of the most able graduates, industry would lose ground both in productivity and technological innovation and hence ultimately jeopardise the country's ability to finance science and education; and increased flow of potential engineers, technologists and scientists from the schools into higher education would also be put at risk.

38. We do not of course wish to suggest that Universities and other institutions of higher education have retained too many graduates but a pattern of employment which denudes industry and the schools of first-class graduates gives cause for concern. It is in part clearly the result of necessary and rapid growth of higher education, and of a comparable expansion of the activities of the Research Councils. To this extent the levelling off of the requirements for teaching staff in the universities could lead to some redress of the balance towards other employment, provided that Research Council policies were adjusted accordingly.

39. Nevertheless patterns have been set and the expectations raised of careers in research which are unlikely to be satisfied in this

country. It cannot be taken for granted that the flow of newly qualified manpower, particularly at the higher degree level, will readily switch to industry and the schools. Employers must make strenuous efforts to make such careers more attractive to able graduates. It seems to us that further pressure to emigrate must inevitably be generated among our best graduates if the present attitudes and patterns continue.

5.11 The Need for Sophistication in Manpower Planning

From the Committee on Manpower Resources for Science and Technology, Report of the 1965 Triennial Manpower Survey of Engineers, Technologists, Scientists and Technical Supporting Staff, Cmnd. 3103, October 1966.

By the mid-1960s more sophisticated forecasting techniques were being used to predict manpower demands.

C. Shortages

47. No analysis of current problems of scientific and technological manpower policy can escape the evidence that there are shortages in specific categories of employment which are not always directly comparable with educational disciplines. Examples of this kind, for which we have seen evidence of shortage, are systems designers and advanced programmers, control systems engineers, and production engineers. These categories are not generally produced by the educational system; they belong to the category of designer or product technologist, which can only effectively be produced by a period of appropriate specialist education and training. Our industrial members similarly stress that the most urgent problem now before us in meeting industrial demands is to devise efficient and attractive methods for making the transition from graduation to employment in specific product groups.

48. The identification of shortages in terms of educational disciplines overlooks the possibility that people of different disciplines may well be suitable for a particular job. Where 'substitutability' is limited, and the educational discipline is a largely sufficient preparation for a first responsible job, the remedy lies in increasing the supply of individuals qualified in that discipline. On the evidence before us, there is likely to be a continuing shortage of mathematicians for the many diverse functions which they are called

upon to perform in employment. In addition there are likely to be shortages in the main engineering disciplines although, as we have noted above, there is evidence that industry is increasingly prepared to recruit scientists for engineering and technological employment. We are studying further the question of substitutability and the implications of particular shortages in relation to specific educational and training requirements.

D. Interpretation of need

49. As we have pointed out in paragraph 35 above, manpower needs are deduced from general economic and technological forecasts. Many depend upon the more effective utilisation of engineering, technological and scientific skills. Some examples are the need for technological viability in our aircraft, shipbuilding, machine-tool and computer industries, for better standards of design and means for the training of larger numbers with design skills, for better utilisation of scientific knowledge of materials, for better standards of mensuration and calibration, and for the more scientific exploitation and conservation of our limited national resources. There are indications that at least one other nation of broadly comparable resources (Japan) is already devoting to these ends larger numbers of qualified manpower than we are. The combined technological resources of the European Economic Community are well ahead of our own. Technological progress in this country, which has depended for so long on an earlier craft-based phase, clearly requires a more extensive and intensive effort towards meeting the needs mentioned above than can be discerned in present trends.

50. Manpower policy acutely requires more sophisticated methods for the estimation of long-term movements of need, and of demand, for qualified manpower. It is hardly practicable to invite the majority of employers to look as far ahead as the decade required to plan developments in higher education; such estimates would involve taking into account general economic and political considerations which they can hardly foresee. Yet manpower planning is essentially a long-term process, insofar as major adaptations to the educational system are concerned.

51. We therefore hope that the capabilities of computable models of the economy can be developed to the stage at which reliable indications can be given of need for qualified manpower. One of the main difficulties must be the establishment of reliable quantitative

relationships between economic performance and manpower utilisation, developed in sufficient detail to throw light on the requirements for manpower at all levels. One of our members, Professor Stone, is engaged in research on this problem.

5.12 What to do about the Brain Drain

From the Committee on Manpower Resources for Science and Technology, Report of the Working Group on Migration, under the chairmanship of Dr F. E. Jones, FRS, *The Brain Drain*, Cmnd. 3417, October 1967.

Concern about the migration of British scientific talent first manifested itself in the early 1960s and thereafter grew in intensity and volume. The Royal Society had sought to measure the extent of the flow in 1963, but it awaited the Jones Committee to examine the 'drain' in depth. Extracts from the Committee's analysis and recommendations are given, but what emerged was that there was no simple solution ready to hand.

9. There is ample evidence that in the eighteenth and nineteenth centuries the United Kingdom, made prosperous by the development and exploitation of its resources and by its native wit, was one of the centres of attraction in this movement. This was before the great engineering and technological explosion accelerated by the 1914 and 1939 world wars, before the emergence of the United States and Russia as two major world powers, and before the advent and rapid development of the aeroplane, and of other quite phenomenal improvements in the technology of automation and communications.

10. There is little doubt, too, that the United Kingdom has been both beneficiary and benefactor in this movement. It has both attracted talent and sent talented people to all corners of the world. Until the Second World War much of this talent went to what is now the Commonwealth, to administer, to defend, to discover and exploit natural resources, to settle. Some of this talent went to the United States, sometimes for much the same purposes, and sometimes to the universities there which were evolving as outstanding research centres. Indeed, there is still a continuing and substantial volume of emigration from the United Kingdom to these places, particularly to the older nations of the Commonwealth

and to the United States. It has remained an accepted part of the whole social and political scene for the British, and especially for young British families, to emigrate and start life afresh in the still-expanding communities of these newer countries. Why then the recent public concern about what has come to be called 'the brain drain'?

11. There is no doubt that public opinion is sensitive to any image of this country which suggests a flight of talent. The 'brain drain' is a term, unknown just a few years ago, which implies that the new selective migration will have a bad effect on the future prosperity of the country. The innate ability of the emigrants may be little different from that of past generations, but nowadays the community is well aware that it has paid heavily for the education and training of those concerned.

15. In considering the movement of qualified manpower as a socio-logical phenomenon, we should as far as possible distinguish between emigration (permanent loss) and temporary migration. In this sense temporary migration may be highly desirable and beneficial. The qualified engineer, technologist or scientist who goes to some overseas university for a postgraduate or post-doctoral course of study in his particular discipline, or enters an advanced overseas industry to gain some additional experience, or takes a course at an overseas business school, will return with additional qualifications and will be more valuable to his country. This also holds true if he has spent the time working in a develop-ing country: he will probably have exercised greater responsibility at an earlier stage in his career, and so have developed more rapidly than might have been possible in the United Kingdom personal qualities of self-reliance and adaptability. Emigration on the other hand is generally undesirable in an economic sense, although it may have desirable aspects as a safety valve for an occasional sur-plus in some particular discipline, or as a means of allowing an individual to use his talents to the full where opportunities do not exist at home. Leaving these desirable factors aside, emigration becomes a disadvantage when it begins to damage the prospects for proper development of the industrial, educational or governmental services of a country.

16. It follows that any international system of migration which has the effect of permitting or encouraging the most highly gifted

of any nation to move to countries with higher standards of living, to the detriment and impoverishment of their own countries, can ultimately do nothing but harm. At the same time, the training of young people to high standards of skill within any one country, unless accompanied by a national ability to provide them with intellectually satisfying and economically worthwhile jobs in which they can use those skills, leads to an unstable situation which can only be resolved by emigration to some other country which can provide a rewarding future. The alternative is frustration, and a quite unjustifiable waste of scarce human talent. There are clearly inherent contradictions in international migration at present which require serious attention.

17. The education of highly skilled engineers, technologists and scientists in any one country does not, of itself, create wealth. Their education must be balanced by Government and industry working in close and detailed collaboration to employ their skills fully to increase the national wealth. The lack of creative national thinking, and of adequate collaboration between Government, industry and educational institutions can be a major contributory factor to the flow of qualified manpower towards the highly developed industrial countries of the world. Moreover, in our view, this criticism can be levelled equally at both recipient and donor countries. It applies to the developing countries whose state of development and political aspirations are often in conflict. It applies to the United Kingdom where employers have not so far created opportunities for the fullest use of this type of manpower. It applies to the United States of America whose eagerness to be at the forefront of the exploration of space and ocean is not matched by any prior coordinated attempt to supply the necessary manpower resources from its own population.

132. This loss of talented manpower has tended to run in cycles corresponding with industrial investment and the general economic climate in the United Kingdom. But the rising trend since 1961 has, we consider, been influenced by new and disturbing factors, such as the attentions of recruiting agents operating on behalf of United States industry. Given the situation of continuing shortage in the United States, and the declared intention of these recruiting agents to step up their activities here, we conclude that the loss of qualified engineers and technologists in particular, and of scientists to a lesser extent, is likely to increase rather than

diminish. This too is an alarming prospect. We come therefore to our first main finding.

> Our investigations confirm that there is already a serious brain drain of talented young engineers, technologists and scientists from the United Kingdom. This is now increasing to such an extent that it constitutes a threat to the interests of this country.

134. It has been suggested in some quarters that one way to curb the brain drain would be deliberately to restrict the movement of those holding specified qualifications. This is, in our view, no answer to the problem for the United Kingdom. It may have superficial attraction as a short-term measure, but to restrict movement where there is no domestic outlet for their skills would both impoverish mankind and frustrate the individuals by causing a waste of their talents. Any such restriction would moreover be inconsistent with the democratic principles by which our society is governed.

> We recommend therefore that the brain drain should not be controlled by imposing restrictions on the freedom of individuals to move across national frontiers.

136. In much the same way, we reject the suggestion, as a counter to the brain drain, that the individual's higher education should in effect be given to him on loan, and that he should be expected to repay it either in cash or by a specified term of service to this country before he is allowed to emigrate. There may be valid reasons, other than those relevant to the brain drain, for such a system of finance for higher education. But as a means of curbing the brain drain it is subject to the same criticism as the idea for national reimbursement discussed in the preceding paragraph. It is also subject to strong criticism on practical grounds. For instance, how would such a control be enforced? Should a graduate in specified disciplines be required to stay for a specified period; or to deposit £x thousand for his passport, to be forfeited if he does not return within a given time? Or should we expect other nations to pursue civil debts on our behalf?

> We therefore conclude that financial controls, whether in terms of reimbursement of one nation by another for the qualified manpower it receives or of financial guarantees by the individual that he will stay in or return to his country of education, are neither

proper nor practicable weapons to be used in countering the brain drain. We recommend that they should not be adopted for this purpose.

140. On the contrary, we consider that if effective and lasting solutions are to be found, they must be sought in firmly positive directions. We consider therefore, that as a positive remedy, the right solution to the brain drain is to create more challenging opportunities in this country, particularly in industry, for young qualified engineers, technologists and scientists in the most creative and productive phases of their career. To achieve this will require a considerable change of attitude in the whole of United Kingdom society. We would agree here with the sentiments expressed in 1964 by Sir George Pickering, Regius Professor of Medicine at the University of Oxford in commenting on the transfer of a research team to the United States. He said:

'*The present instance is symptomatic of the relative poverty of our society. This relative poverty will only be overcome when we as a nation become once more interested in the creation of wealth rather than its distribution and consumption.*' (*The Times*: 11 February 1964.)

141. It is our belief that successive Governments have, over the years, given undue encouragement and emphasis to the policy implied in this quotation. At times, of course, governments have been forced into this by weight of economic and political circumstances, but the undoubted result has been to lay insufficient emphasis on the innovation and productivity which are so vital to the economic well-being of the country and therefore to its capacity to attract, utilise and retain the best engineers, technologists and scientists who in their turn are the foundation of the next round of advance. Unless the economy is strong these people will not be attracted, and if they are not attracted the economy cannot remain strong in a highly competitive world.

We consider, therefore, that there is insufficient recognition, at all levels of our society, of the fact that the source of national wealth is in the creative, productive industries of this country. It is here that the engineer, technologist or scientist has both the greatest contribution to make and the greatest challenge to face. We recommend that Government and Industry should combine to emphasise that new ideas alone do not create national wealth and prosperity, and that they contribute to national prosperity only when they are vigorously exploited in productive industry.

5.13 Problems and Dangers

From the Committee on Manpower Resources for Science and Technology, Report of the Working Group on Manpower for Scientific Growth, under the chairmanship of Professor M. Swann, FRS, FRSE, *The Flow into Employment of Scientists, Engineers and Technologists*, Cmnd. 3760, September 1968.

110. We see these problems and dangers in the present situation:

I A flow of qualified manpower into the schools inadequate for their needs; that is for the future needs of the individual pupil, of science, and of the economy. Teaching is vital for the growth of science and technology and must itself be adequately nourished. It must become competitive in order to attract the most able graduates. This is one essential factor in removing the present threat to the long-term supply of scientists and technologists.

II A threat to economic growth and technological advance in industry, unless there is a greater injection of scientists and technologists and particularly of the most able graduates; industry must recognise and exploit the changing pattern of supply, and must devise new ways of using science graduates, especially those with Ph.D.s, not only in specialist posts, such as research and development, but more widely in manufacturing and production, and in administrative and managerial positions.

III A first degree course in science and technology that, while providing a growing proportion of new supply, is generally not sufficiently widely based to meet either the desire for breadth in schools or the longer term needs of the individual and his employer – whether it be for successive phases of specialisation required by technological change throughout the career or for employment in more general functions.

IV An increasing tendency towards postgraduate education that is not, in general, closely geared to the needs of the economy. A strong preference for research training on a scale unlikely to be satisfied by employment opportunities over the next five years, and that could lead to increased emigration and dissatisfaction with careers in science and technology, affecting disproportionately the career decisions of the next generation, unless present patterns of deployment and utilisation are changed.

V Inadequate mobility of qualified manpower both in nature and scale; greater mobility is needed to respond flexibly to technological change and to achieve the redeployment necessary in an increasingly science-based economy; and to avoid both a consequent risk of isolation of sectors that are interdependent and undue reliance on new supply as a means of changing reactionary patterns and predilections.

6 The Scientific Civil Service

In 1969 there were about 470,000 non-industrial civil servants in the United Kingdom. The main classes of employment with their approximate strengths were:

Administrative Class	3,900
Executive Classes	80,000
Specialist Classes	130,000
Clerical and Ancillary Clerical Classes	217,000
Messengerial Classes	26,000

The Specialist Classes include the Professional and Technical Classes as well as the Scientific Class and cover a very wide range of activities: accountants, architects, lawyers and surveyors, in addition to scientists and engineers.

Technically qualified men have been employed by the government since at least the middle of the last century, as we indicated in the Introduction. They were 'experts' in the sense of being men with special knowledge and skills and thus, by extension, essentially different from the generalists, the administrators and executives of the Civil Service.

Attempts to systematise the arrangements for the employment of those scientists and technicians in government service date from the first world war. Even so, until 1930, when the report of the Carpenter Committee was published,* the situation was highly anomalous; within the Professional, Scientific and Technical Classes there were more than five hundred distinct grades although these classes at that time comprised less than 10 per cent of the non-industrial Civil Service. The Carpenter Report led to the establishing of a two-tier structure of a Scientific Officer Class and a Scientific Assistant Class.

But there were other anomalies too, and in the 1930s it was possible to write: 'The scientist in Government pay gets the worst

* H. M. Treasury, Committee on the Staffs of Government Scientific Establishments.

of both worlds. He does not have academic privileges and he misses the possibilities of advancement and in many cases even the security of tenure of the Civil Servant.'*

The second world war brought about a re-evaluation of the role of the scientist and this was embodied in the report of the Barlow Committee in 1945 [6.5] in consequence of which a three-tier system was introduced. The innovation was the formation of an Experimental Officer Class as a strengthened version of the Scientific Assistant Class. Barlow sought to produce a unified Scientific Civil Service with common classes, conditions of service, grades and salary scales with adequate scope for mobility between departments and thus to attract a better quality of entrants than had often been the case between the wars.

Further suggestions with regard to the restructuring of the Scientific Civil Service were investigated in 1964–5 by a Treasury-appointed committee under Sir Mark Tennant [6.9]; amongst other things this looked at a proposal to merge the Scientific Officer and Experimental Officer Class but eventually decided against it, primarily on the grounds that it was desirable to preserve the prestige of the Scientific Officer Class. The Institution of Professional Civil Servants later proposed to the Fulton Committee on the Civil Service [6.10] the setting up of a Technology Group, but by mid-1972 no action had been taken on any of these suggestions.

Questions of status and security are, administratively, relatively easily solved and to a large degree many of the Service's earlier battles are now over. But the other major concern throughout our period, that stemming from the Service's 'expert' function, has not proved so tractable since it has depended to a greater extent on less tangible factors and, perhaps prime among them, a particular attitude of mind.

An indication of the light in which the scientists and technologists are viewed has been seen in the low numbers of scientifically or technologically trained recruits to the Administrative Class in the Home Civil Service; this was 3·4 per cent between 1948 and 1956 and had risen only to 4·5 per cent between 1957 and 1963. Furthermore the Scientific Civil Service for long felt itself excluded from its rightful place in policy formulation and implementation [6.1, 6.3, 6.4].

This attitude of mind has been succinctly enshrined in Sir Walter Elliot's words that the scientists should be 'on tap but not on top', the reason being that 'the skills of the generalist are valuable at the top of government because those without specialist biases can best

---

* J. D. Bernal, *The Social Function of Science* (Routledge, London, 1939), p. 107.

fit together contributions of experts in many areas'.* It is not easy to see either a justification of this attitude [6.6] or a simple means of changing it [6.10], in spite of what was referred to in the Fulton Report as the Scientific Officer Class's 'highly privileged position' [6.10].

6.1 Utilisation of Science in Public Departments

From *British Science Guild Journal*, June 1921, no. 13, pp. 33–5.

In February 1920 the British Science Guild set up a Committee on the Utilisation of Science in Public Departments under the Chairmanship of Dr John W. Evans, FRS, of the Imperial Mineral Resources Bureau. The *Journal* article summarised the proceedings and findings of this Committee.

The Committee has been forced to secure the attainment of the following aims:

I. That in the national interest, science shall be adequately represented in public departments.

II. That full weight shall be given to the opinion of the technical officers of the departments whenever decisions have to be made which depend mainly on scientific factors.

III. That whenever circumstances render it desirable, reference shall be made to the highest technical authorities outside the departments for their advice and assistance.

IV. That adequate remuneration shall be given for expert advice and assistance afforded to Departmental Committees or otherwise in the public service.

V. That there shall be an adequate scale of remuneration for those employed in scientific work in public departments, and satisfactory conditions of service, especially of promotion and superannuation, of those permanently so employed.

VI. That whenever desirable, public departments shall secure the cooperation in their work of the Scientific Staff of Universities and Technical Colleges.

* R. Rose, *Politics in England* (Faber, London, 1965), p. 100.

VII. That the regulations for admission to scientific posts in public departments and to the higher Civil Service generally shall be so framed, in consultation with the Universities and Technical Colleges, as to secure that those appointed to the former have received an adequate training, and that the latter posts include at least as many graduates in science as in arts.

The members of the Committee feel ... that they ought to express very strongly their opinion that if the value of science is to be realised as it should be in public departments, measures should be taken to secure that scientific knowledge should be given adequate weight in the competitive examinations for administrative posts in the Civil Service. At present, in spite of recent changes, this is far from being the case. It would still seem to be to the advantage of a prospective candidate to specialise in literary subjects which do not require the time and expense involved in laboratory work. It would be very desirable that the regulations should be modified sufficiently to ensure that at least half the successful candidates should have given serious attention to science. It is difficult to exaggerate the value of an infusion among the administrative staff of a certain proportion of men and women who have a genuine love for science and have even in some cases engaged in scientific research.

6.2 Secrecy in Research

From Report of the Committee on the Staffs of Government Scientific Establishments, under the chairmanship of Professor H. C. H. Carpenter, FRS, H.M. Treasury, 1930, p. 37.

The Committee was set up in May 1929 to examine the functions and organisation of the establishments. Although largely concerned with matters of recruitment and the structure of the Service it also looked at other aspects of work in government service including the controversial matter of secrecy. This has traditionally been held to be alien to the spirit of science and those scientists who submit to arrangements intended to preserve secrecy have been described by Shils as rendering themselves into 'epiphenomena of the scientific community. They take but they do not return anything. . . .'*

* *Advancement of Science*, June 1968, p. 473.

Our attention has been drawn to the condition of secrecy imposed on the work of some scientific and technical officers. The strictness with which this condition has to be observed varies considerably between departments; it is practically absent in departments such as the Department of Scientific and Industrial Research, where the work has close affinity with outside industry, and is most onerous in branches of the Fighting Services, where such officers are engaged in research on the construction and use of modern weapons of warfare. The necessity for this secrecy, in its varying degrees, is a matter of policy to be decided by the departmental authorities; nevertheless we cannot omit to note that it imposes a real disability on the scientific staff. A research worker can only achieve recognition in the scientific world or secure the highest academic distinctions if additions to knowledge discovered in the course of his researches can be published for the consideration of all concerned. Given the necessity for secrecy, the mitigation of this disability does not lie in action which can be taken by the Government alone, since the cooperation of the Universities, and possibly of the scientific societies also, is clearly required. For this reason we have not felt justified in pursuing the question with a view to making any recommendation about it. We think, however, that the three Fighting Services might explore together the possibility of asking the Universities, or perhaps one University only, to consider the revival of an arrangement which we believe was at one time in force, namely, the acceptance of secret theses for scientific degrees.

6.3 Machinery of Government in Relation to Modern Requirements

From a Statement submitted by the Association of Scientific Workers to the Royal Commission of 1929–30 on the Civil Service (the Tomlin Commission). Minutes of Evidence, 43rd Day, 12 November 1930.

43. We have no desire to depreciate the value of classical and other so-called humane studies. We acknowledge the debt which our civilisation owes to the classical Greek philosophers. We appreciate the importance of Latin in former centuries as the vehicle for the spread of knowledge throughout Europe. We should expect to find little antagonism between the administrative and 'expert' elements

in the State Service, if the study of classical Greek had the effect of inculcating an appreciation of the unique contribution of the ancient Greek philosophers to world progress, the essential of which was the scientific study of the external world, a study which gave realism to every manifestation of Greek culture: or if the student of Latin recognised that knowledge of that language while auxiliary to the study of history and law, and useful in the study of modern languages, was no adequate compensation for ignorance of modern scientific thought and development and twentieth-century economics.

44. The late Lord Haldane expressed the opinion that most of the friction between administrative and scientific experts is due to lack of appreciative understanding of each other's functions, not infrequently because their respective functions are ill-defined. There is a tendency for the purely administrative elements to encroach on the domain of the experts, leading inevitably to the demand of the experts for freedom from what they suggest is ignorant administrative interference. It has been suggested that the solution to this vexed problem is the business of the political heads of government departments: that they should select as administrators civil servants capable of appreciating the limits of their own knowledge and the extent of their need for guidance from experts. This, of course, presupposes the existence of political heads of departments with a broad conception of their ministerial functions, knowledge of the needs and aspirations of the people they represent, more than a casual acquaintance with the trend of expert opinion bearing on their duties, and a flair for the selection of the right personnel. It assumes also that there exists in the administrative grades of the Civil Service a sufficient number of men and women with a knowledge of science.

45. It is clear, however ... that the number of administrators with a scientific training is inadequate to the needs of a modern State. It is also true that few political heads of departments have any knowledge of science, and what makes the position worse is that the scientific experts in the State Service rarely have access to or otherwise come in contact with the political heads of departments. The scientific experts have no guarantee they will be consulted by the administrative heads of their departments even when it is necessary that they should be, and decisions upon matters within their legitimate sphere of activities and affecting their

interests may be taken without their foreknowledge. The consequence is that scientific workers in the service of the State feel that they constitute an element alien in spirit to the system which has absorbed them, a system which to their mind is too inelastic to make proper use of the knowledge which the State has at its disposal.

6.4 The Expert in the Civil Service

From F. A. A. Menzler, in *The British Civil Servant*, edited by W. A. Robson (Allen & Unwin, London, 1937), pp. 163–85.

The Tomlin Commission [see 6.3] brought about no radical restructuring of the Scientific and Technical Classes of the Civil Service. Menzler, a former chairman of the Council of the Institution of Professional Civil Servants and a vehement critic of the Tomlin Report, discussed, among other topics, the opportunities accorded to the expert in the Service in the framing of policy.

Most students of public administration are familiar with the manner in which the typical Whitehall department works. At the apex of the pyramid is the Minister with whom, subject to the over-riding authority of the Cabinet and Parliament, decision upon the action to be taken or the policy to be followed rests. Ordinarily, his immediate adviser is an administrative expert – not necessarily a technical expert – known as the Permanent Secretary, supported by a phalanx of members of the Administrative Class whose job it is in theory and generally in practice to do the staff work of administration so that at the critical moment the Minister may be advised in lucidly drafted memoranda upon the pros and cons of the alternative policies from which he may choose. It is even said that on occasion the Minister is advised to adopt a particular policy as against other alternatives that might be advocated. The Minister is, of course, free to accept or reject the advice tendered to him. Once the decision is taken it is the tradition of our non-political Civil Service that the Minister shall be faithfully served, whatever his political colour, and however wrongheaded and repugnant to those who really know the facts the policy he decides upon may be. Where does the technical expert come in? Many ministerial decisions in a society which is making ever-increasing use of scientific discoveries depend upon technical considerations. The question naturally arises as to what extent the expert is permitted to take his share in the formation of policy and, perhaps even more important, to what extent he is allowed by ordinary Civil Service practice to see the Minister and, in person, explain to him the reasons for the technical advice he may be called upon to render.

It is suggested that prima facie evidence of an unsatisfactory state of affairs from the point of view of the intervention of the expert is afforded by the traditional structure of Whitehall administration itself which, in outline, has been described above. It cannot be denied that if some matter of public importance arises upon which the Minister desires advice, he will ordinarily go to the secretariat, for the simple reason that that is the machine. The secretariat will collect the facts and, if technical considerations are involved, may conceivably, but not necessarily, seek the views of the technical experts in the department, if any. The experts will draw up those memoranda so characteristic of Civil Service procedure, which will be forwarded to the appropriate member of the Administrative Class, possibly an Assistant Secretary but sometimes no more than a Principal, who will do his best to understand what it is all about; and in the light of such appreciation of the technical factors as he is able to attain (and sometimes the technical experts are not very helpful in this respect) and of non-technical factors of a political or administrative type, he will frequently draft another suitably peptonised report or minute or memorandum for the consideration of perhaps a Principal Assistant Secretary or even of the Permanent Secretary himself. After these successive processes of filtration the Minister gets something before him upon which to reach a decision.

For obvious reasons it is impossible in the nature of things to quote individuals to demonstrate that on this or that occasion their advice was ignored, misconstrued, perverted or flouted, or that they were refused access to the Minister. Again, evidence of particular cases can always be countered by other particular cases, for example, where a technical man of outstanding personality has by virtue of sheer force of character ensured a hearing for the policies he has advocated. Let us assume, however, that the Minister takes a decision which does not give full weight to or ignores some relevant technical consideration. The technical expert, if he hears about it in time and is senior enough and is of outstanding personality, may succeed by sheer force of character in getting to the Minister and explaining, in person, the technical aspects of the matter. But the number of men, whether expert or non-expert, with the resounding type of personality which nobody can ignore, is extremely small. We may push on one side any arguments in defence of existing practice which are based upon the fact that particular experts have had, in fact, a profound influence on public administration in spite of the handicap of being expert in the subject-matter under consideration. When all is said and done, it is the structure of the organisation that matters and not the manner in which that organisation may be made to work as a consequence of the idiosyncrasies of particular individuals. A system of organisation that is dependent for its success upon the manner in which the organisation is ignored is fundamentally unsound and cannot conduce to efficiency in the long run.

Against this account of the process of administration must be set the fact that particular Ministers, owing to their scientific training or special interests, have been known to go direct to the technical advisers in the departments, much, of course, to the disgust of the administrators. But for the reasons already indicated, this does not make the system a sound one. It is the normal functioning of the machine with which we are concerned and not with the eccentricities in its working. Again, a defender of the existing system would say, as, indeed, the present Permanent Secretary to the Treasury told the Tomlin Commission, that in practice the grievance – he described it as an 'amiable grievance' – is non-existent because nobody would be fool enough not to avail himself of the advice of the expert in arriving at decisions on policy. This, it may be respectfully suggested, is not the point. The issue is the effective part that the expert is permitted to play in the formation of policy, as his normal function under the organisation. . . .

It would not be fitting to conclude this broad survey of the position of the expert without an emphatic disclaimer that there is in the foregoing any suggestion of a deliberate policy aiming at keeping the expert in subjection. His position in the Civil Service is evidence of something rooted far more deeply than Treasury Chambers. Its source is to be found rather in the attitude of society itself to those who in so large measure are responsible for the form which civilisation is taking. Although our statesmen will in their moments of relaxation tell us that this is a scientific civilisation, few of them are prepared to follow out in word or in deed the social consequences of this self-evident fact. The *malaise* of present-day society arises not so much from the shortcomings of politicians as from a general failure due to mis-education to realise that scientific research and its foster-brother technical advance are producing strains and stresses in the structure of society which, if they do not receive prompt attention, may bring about the destruction of the social order. The status of the specialist in the Civil Service, which to the casual observer may seem no more than one of the peculiarities of the service, is profoundly symptomatic of society's attitude towards those who are reshaping, with ever-increasing velocity, the social environment.

6.5 General Conditions in the Postwar Service

From Report of the Barlow Committee on Scientific Staff, 1943. Printed as an Annex to Cmd. 6679 of 1945: *The Scientific Civil Service, Reorganisation and Recruitment during the Reconstruction Period.*

In 1946, the Scientific Classes of the Civil Service were reorganised; the highest tier of the new system, that of the Scientific Officer Class, assumed responsibility for matters of scientific research, advice and developments within the Civil Service. This reorganisation followed the recommendations of the Barlow Committee in 1943 which concluded that at least part of the government's difficulties in attracting the best scientists lay in the conditions of service.

15. *General Conditions of Service.* – One of the dangers to which in the nature of things a Government Scientific Branch is liable is the tendency towards isolation from the rest of the scientific world. This not merely has an effect upon the value of the work of such Branches, but acts as a hindrance to recruitment. Young scientists, ambitious to obtain recognition for their work and to keep abreast of new developments, are discouraged from embarking upon a career which appears to remove them from contact with the Universities, with learned societies and with the research side of industry. The prospect of a closer relationship with the outside scientific world would remove one of the impediments which weigh with the University authorities in the advice which they offer to their students on the choice of a career.

L

This 'isolationist' tendency is one which in our view should be eliminated to the greatest practicable extent. The following are expedients which may be helpful to that end:

(*a*) The development of extra-mural research contacts between Government Departments and the Universities over the past few years has proved of great benefit. Heads of Scientific Departments should endeavour constantly to improve and strengthen those contacts. They should take every step to interest both potential recruits at the Universities and the authorities in the work of their Departments, whether by lectures or in more informal ways.

(*b*) An attempt should be made to develop a system of temporary interchanges of staff, particularly during the early stages of their career, between Government Departments and the Universities. One effect of such a scheme would be to impart new vigour into the work of the Government establishments.

(*c*) The Committee attach great importance to the introduction of a scheme on the lines of a 'Sabbatical Year' under which selected scientific staff would be allowed a period or periods of special leave to go and work at a University or other institution at home or abroad at the expense of the Department which employs them.

(*d*) Departments should be empowered and encouraged to give staff special leave (with pay) to enable them to attend meetings and conferences of a scientific character.

(*e*) Similar close contacts with industry would be almost as desirable, although here certain difficulties are bound to arise. Firms will be anxious that trade secrets should not be divulged, and Government Departments will naturally wish to avoid not only the possible leakage of official secrets but also the appearance of favouring one firm as against another in the form of close contact with Government work. No doubt, there are other difficulties such as the fear of losing staff seconded to industry (though this might be lessened by an improvement in the conditions of service such as we suggest), or again the reluctance to spare good men for a purpose which will not necessarily bring immediate advantages, however much it may prove beneficial in the long run. We are not convinced that these difficulties are insurmountable. If they can be overcome, the Government Service would benefit from such exchanges.

(*f*) It should go without saying that Government Scientific Branches should keep in the closest contact with each other. The contact is, we believe, already close on programmes of research, but it has been suggested to us that in matters of administration, and in particular staffing questions, liaison is too often effected indirectly through the central Establishment Branches of the Departments concerned. Central control of staffing matters by the Principal Establishment Officer of a Ministry is a fundamental feature of the Public Service, but methods should be evolved to bring scientific 'administrators' in touch with each other without impairing that control.

(*g*) While we would not go so far as to recommend a unified State Scientific Service with a single Establishment and responsible to a common head, transfers of staff from one Department to another appear to us to be very desirable. Such transfers have become increasingly frequent in other fields in the Civil Service with good results. The effect, so far as individual scientists of ability are concerned, would be to enlarge the field of possible promotion; there might be a corresponding diminution in the prospects of 'sitting tenants', but this would have to be faced. We consider that the advantages of a scheme of relatively free transfer would outweigh the disadvantages.

(*h*) It has also been suggested to us that a defect in organisation in certain Departments is that an individual scientist is not able to follow through to its conclusion the development of a particular line of research which he has initiated. We regard it as most important that wherever it is practicable (and it obviously is not always so) researchers, and in particular the younger men, should when personally suitable remain associated with the development of projects which they themselves started. We know that the Directors of Scientific Research whom we have consulted appreciate the value of the encouragement which can be given in this way, but its importance merits more general acceptance.

(*i*) An important point in regard to research establishments, which may easily be overlooked, is their geographical location. It would be of great advantage if research stations could be situated within easy reach of a University or other centre of intellectual life. Moreover, it would be desirable to secure close proximity between different establishments working on kindred research problems. It should also be made easy for individuals at any level in the hierarchy to make direct contact with their opposite numbers in other establishments.

(*j*) One of the greatest disadvantages of Government Scientific Service is the extreme emphasis upon secrecy. This has, we believe, been a powerful factor in hindering recruitment. It operates adversely in two ways. It makes it difficult or impossible for the Government worker to enter into discussion with similar workers outside, and it prevents his obtaining recognition in the scientific world of the value of his work. It must be remembered that scientific ideas are rarely the product of a single brain; they arise from the contact and clash of separate minds. Both flint and steel are needed to produce the spark. If the Government condemns its scientific workers to a life of monastic seclusion, it will inevitably suffer from a paucity of new ideas. We recognise that the relaxation of secrecy restrictions is a difficult matter, particularly in relation to the work of the Defence Departments. Nevertheless it is most important that they should be relaxed, and we recommend as a possible solution an arrangement on the following lines:

(i) It should not be assumed as a matter of course, at any rate in peacetime, that the whole of the work of a Research Establishment must automatically be regarded as secret;

(ii) Matters of fundamental research should be regarded as prima facie suitable for discussion with outsiders;

(iii) The questions on which absolute secrecy is prescribed (such as the development of particular secret devices) should be kept to the minimum;

(iv) If, as we hope, there is to be a greater liaison with outside institutions, the head of a Research Establishment should have discretion, and indeed be encouraged, to allow his staff to discuss their work confidentially with any members of those institutions who are engaged on similar problems;

(v) A system should be possible whereby members of the staff engaged on secret work could submit theses for higher degrees. Certain Universities already provide for this.

(*k*) We think that fuller use might be made of Advisory Councils. It should not be enough that they should meet three or four times a year to consider a miscellaneous agenda. Their members should be encouraged and expected to visit the laboratories and discuss the work with the staff. We believe that the effect on the staff of such discussions and of the feeling that an eminent man from outside knew what they were doing would be most stimulating.

Alternatively, the method might be tried of appointing either a single paid consultant or a panel of specialists who would give a substantial amount of time to visiting the laboratory and talking to the staff. This would not preclude an Advisory Council concerned mainly with the general programme and policy of the institution.

6.6 A Scientist in and out of the Civil Service

From Sir Henry Tizard's Haldane Memorial Lecture, delivered at Birkbeck College, London, 9 March 1955, pp. 18–20.

Tizard left the Civil Service in 1929 for a long spell in university life. He rejoined the Service on a fulltime basis in 1946, as Chairman of both the Defence Policy Research Committee and the Advisory Council on Scientific Policy. In his Haldane Lecture he summed up his reflections on the role of the scientist in the Civil Service.

A much more serious difficulty is caused by the deplorable intellectual gap that exists between those who have had a scientific education and those who have not. Practically all Ministers and members of the administrative civil service belong to the latter class. Every scientist who has been concerned directly or indirectly with the formulation or execution of policy must have experienced the baffling difficulty of conveying his thoughts to someone who does not know the a b c of the subject. It is different with economics. Many civil servants have had a university education in economics, and even a classically educated Treasury official acquires much knowledge of economics in the ordinary course of his business. He can argue intelligently with economists on their own ground, and can appreciate the reasons for their advice even if he deprecates the jargon in which they are clothed. I do not want to make too much of this, or to imply that there is a lack of sympathy between scientists and administrators. There is not, as a rule; but there is a lack of understanding, which will continue until far more administrators have had a scientific education. At present the number of scientifically educated men entering the administrative civil service by competition averages one a year, which is not enough. No criticism of the Civil Service Commission is implied here: I simply state a fact.

One way of securing more scientific administrators would be to take more trouble to implement the recommendation of the Barlow Committee of 1945 that suitable members of the scientific Civil Service should be transferred to the administrative class. This recommendation has had little effect, so far. There is no particular temptation for a young scientist to seek a transfer. On the other hand, I think that there are many scientists in experimental stations who have done their best scientific work by the age of forty, and would welcome a transfer to a post of equal status in the administrative class if they could be tolerably certain that they had the necessary all-round ability and interest in the work. It is no good throwing such men into the deep end of administration only to meet an intellectual death by drowning in a sea of paper. They must learn to swim first. There is much to be said in favour of forming a central scientific staff, engaged only on studying questions of policy, under the immediate direction of the Advisory Council, to which young scientists could be seconded for a year or two from the Research Establishments. It would soon be discovered if they had the necessary qualities for higher administration. If they had, they could be marked for transfer at a later date if they wished: if they had not, no harm would be done. They could go back with relief to their normal scientific work. The difficulty about this suggestion is that the only men who would be worth transferring temporarily to a central staff would be men whom the Directors of experimental stations would say that they could not spare – so it would not be popular. But I have had experience which convinces me that it would be valuable.

There still remains the problem of ensuring that the majority of our future administrators, if not all, shall have had a broad scientific education. The fact that most arts students at universities have now had an education in the natural sciences up to a low prescribed standard while they were at school – or so I am told – does not meet the need at all. Much more than that is needed for the higher education of administrators.

6.7 Problems of Staff Management

From Office of the Minister for Science, *The Management and Control of Research and Development*, 1961.

As part of the report the workings of the Scientific Civil Service were also examined [cf. 4.2.3].

Objectives

279. The problems which are associated with the existence of the Scientific Officer Class in the Civil Service have, we think, tended to be confused too much by comparisons between the Scientific Officer Class and the Administrative Class. As a result, too little attention has been paid to the fact that, from the point of view of staffing, there will always be important differences between scientists engaged in research and development and administrators who deal with the more general affairs of a Government Department. One main difference is that, while a proportion of research scientists remain productive either in one or more specialised fields of research for the greater part of their careers, most scientists do their best research early in their careers. On the other hand, the quality of administration is something which is expected to improve with age and experience. A second main difference is that in most types of research, once a man has established himself, there is neither need nor justification for anything like the same measure of supervision or of reference upwards as is required in the Administrative Class. This second difference is reflected in the lower ratio of senior to junior posts in the Scientific Officer Class as compared with the Administrative Class.

280. But these seem to us to be differences which also mean that the intrinsic importance of the work done at an early age by research workers should be appropriately recognised in terms of pay and promotion; and that, in order to hold down the average age of research workers in general, and to provide adequate outlets for older men, deliberate provision should be made for transfers at appropriate stages either to other types of work or to other classes in the Civil Service, or to industry or education. This last point raises fundamental issues to which we return later in this Chapter.

282. Whatever may have been the merits of the recruitment arrangements and conditions of service of the Scientific Civil

Service when it was formed some sixteen years ago, we doubt whether the present structure of the Service or the manner in which it has been administered enables these objectives to be attained.

283. The system introduced in 1945 has resulted in the following age and grade grouping:—

TABLE VI

Grade, salary and age structure of the Scientific Officer Class (Grade and age structure based on information collected in 1958 – see Note (*) below)

Grade	Current Salary £	Percentage of total	Lowest age (years)	Median age (years)
SPSO and above	2,650 to 7,000	20	33	49
PSO	1,716—2,418	39	30	42
SSO	1,342—1,654	29	25	32
SO	738—1,222	12	20	26

Notes (*) Information for the whole of the Scientific Officer Class (including temporary staff) has not been compiled since 1958. The information presented above is, however, consistent with that recently obtained by the Committee from a few Departments, including DSIR, and broadly represents the structure of the Class as it is today.

This distribution is very different from that prevailing in the laboratories of industry, the universities and technical colleges, and the Medical Research Council. The principal defects which it reflects are that too many young scientists are committed prematurely to a permanent career in Government research and, second, that those in their forties and fifties have no option but to stay on as research workers (usually at Principal Scientific Officer level) long after they have made their main contribution to research, and in a period of their careers when their experience and ability would be of value if directed to other activities.

284. Another major criticism levelled at the administration of, and conditions of service in, the Scientific Officer Class by many of those with whom we discussed staffing arrangements was lack of mobility. According to our evidence, this reveals itself in two ways.

First, there is insufficient movement of research staff between establishments or, indeed, from one line of research to another. As a result, Government scientists are deprived of the stimulus of new work, at the same time as they are denied the wider experience necessary for those who ultimately rise to higher positions of responsibility. Second, there appears to be an absence of any planned arrangements for exchange or secondment of research staff between Government research establishments on the one hand, and the universities or industry on the other, or even, indeed, with other branches of the Civil Service. Because of this, members of the Scientific Civil Service enjoy fewer opportunities than they should of broadening their experience in a way which would be beneficial both to themselves and to the Government.

285. We do not wish to give the impression that failure to deal with the problem of mobility of research scientists is restricted to the Civil Service. In varying degrees our criticisms apply also to the research departments of industry and to the universities. Industry is better placed, however, for dealing with the problem, since its research staff have outlets to other activities within their organisations, for example, design, production and sales; and in the universities there is always ample scope for increased teaching or administrative responsibilities.

6.8 Labour's Plans for the Civil Service

From R. H. S. Crossman, MP, 'Scientists in Whitehall', *Encounter*, July 1964, pp. 3–10.

The views of the shadow Minister for Science expressed shortly before the general election of 1964.

But when all is said and done, the sections of Whitehall where the scientist is given a hearing are extremely limited. There are many Ministries which need scientific advisers on their Establishment that have not got them, and there are many more where the anti-scientific attitude of the higher Civil Servants prevents the Minister from really understanding the nature of a problem before he comes to his decision. . . . Nevertheless, to talk as though all we needed in order to put Whitehall right is to inject a lot of natural scientists into key positions in the Ministries is to disregard what is really wrong. The more rapid the rate of technological change to which

Government is subjected, the more urgent will be the need to maintain a permanent cadre of professional administrators who can combine very high intellectual calibre with absolute incorruptibility. I would hope that when we get the balance between Arts and Sciences in our schools and universities right, we shall be able to achieve not only a fair balance between scientists and humanists at the top of the Civil Service, but also a new kind of Civil Servant whose general education is both scientific and humane. Meanwhile, we must make do with the cadres we have. The one thing which would certainly undermine the morale and the efficiency of our Civil Service would be the fear that a Labour Government intended to replace a large number of our senior permanent administrators with scientific experts of its own choosing. By doing this, we should destroy the morale of the Civil Service without any assurance of improving its efficiency. For what is wrong today is not merely the lack of natural scientists in Whitehall but the fact that specialist knowledge of all kinds – whether of natural science or of social science or even of human psychology – fails to get through to the politicians in time to influence the decisions that are being made.

If we are to make Whitehall responsive to technological and social change, the problem we have to solve is how to marry a permanent Civil Service with outside expertise. . . .

But if our aim is to ensure that all government policy is science-based, and that Ministers make their decisions upon the very best information available, it is not sufficient merely to increase the number of whole-time scientists working in government departments. We must also stimulate a steady movement between government and the universities, the universities and industry, and industry and government, so as to ensure that the rigid procedures of the professional administrator are rendered flexible by constant contact with outside realities. (Although on much too small a scale, a circulation of this kind has already been achieved in the Ministry of Defence: we must extend it to every area of government activity; and we can be sure that everybody will be a gainer.) The vitality of our academic research will undoubtedly be increased if leading university scientists, social scientists, economists, and administrators, as a matter of regular routine, spend two or three years on contract to the government in Whitehall, and perhaps another period of two or three years working in industry as well. The same is true of industry. Here, too, new links must be created with government on the one side and the universities on the other, and the best way to do this is by ensuring a regular interchange of personnel.

A sharp increase in the size and in the status of our scientific Civil Service; a full recognition of the vital role of the outside specialist on temporary assignment to Whitehall – these two measures should provide us with the manpower required to ensure that Government adapts its thought and its procedures to technological change, that 'planning is science-based', and that Cabinet decisions are arrived at on the basis of a scientifically-assessed intelligence.

6.9 The Case for Merging the Scientific Officer and the Experimental Officer Class

From the Report of a Committee of Review under the chairmanship of Sir Mark Tennant, *The Organisation of the Scientific Civil Service*, H.M. Treasury, 1965, pp. 9–10.

The Committee were appointed in 1964 to review the organisation of the Scientific Civil Service, including *inter alia* whether there was justification for taking further the Barlow recommendations of 1943 and merging the two higher classes of the Service.

19. It has been suggested to us that the Scientific Officer and Experimental Officer Classes should be merged into one class. It is argued that a reorganisation on these lines could make for greater flexibility, more particularly in the manning and complementing of Research Establishments; also that the image of the Scientific Service might be improved, because the scope for advancement on merit would seem greater in a class where there was a single promotion ladder. Moreover increasing numbers of graduates are now entering the Experimental Officer Class. A unified service might make it easier to make full and proper use of this graduate talent, and might also reduce the effect of the apparent tendency on the part of some universities to discourage graduate entry to the Experimental Officer Class.

20. On the other hand there must be considerable doubt whether a unified class would succeed in attracting the same number of good scientists as does the present system, and more

particularly whether it would attract the really outstanding men. There would no longer be a Class with a special prestige of its own, and we believe that the prestige which attaches to the Scientific Officer Class does, in fact, make an appeal to the right kind of recruit. From the practical point of view the system could complicate the question of probation. At present each probationer is judged on his potential for his present class, but within a unified and considerably more extensive class, a suitable standard of judgement might prove difficult to establish and apply. The system would also call for a more precise definition of the work appropriate to each of the many grades in the class than is necessary under the present arrangements. This could produce undesirable rigidity. It seems likely moreover that, in practice, two distinct streams would emerge, and that when a particular post was to be filled, it would normally be clear which of the two types of officer (speaking in terms of the present structure) could more appropriately fill it. It would thus still be necessary when making promotions to distinguish (whether formally or informally) between the two types of ability and qualification. Alternatively, there would be real danger that undue emphasis would come to be placed on seniority as a qualification for promotion.

21. It is not easy to assess the relative weight of these advantages and disadvantages. Even had we believed that the balance came down in favour of a unified class we should not have felt it appropriate, in view of the very wide implications of such a change, to do more than recommend that the idea should be further studied. But in fact we are satisfied that in present circumstances the balance of advantage lies in favour of retaining the present arrangements. The recommendations which we shall make later would not make it more difficult to unify the two classes if at any time this were considered to be desirable.

6.10 A Way forward for the Service

From the Report of the Committee on the Civil Service under the chairmanship of Lord Fulton, Cmnd. 3638, 1968.

A. Summary of main findings (vol. 1, p. 104–5).

1. The Home Civil Service today is still fundamentally the product of the nineteenth-century philosophy of the Northcote-Trevelyan

Report. The problems it faces are those of the second half of the twentieth century. In spite of its many strengths, it is inadequate in six main respects for the most efficient discharge of the present and prospective responsibilities of government:

(a) It is still too much based on the philosophy of the amateur (or 'generalist' or 'all-rounder'). This is most evident in the Administrative Class, which holds the dominant position in the Service.

(b) The present system of classes in the Service (there are over 1,400, each for the most part with its own separate pay and career structure) seriously impedes its work.

(c) Scientists, engineers and members of other specialist classes are frequently given neither the full responsibilities and opportunities nor the corresponding authority they ought to have.

(d) Too few civil servants are skilled managers.

(e) There is not enough contact between the Service and the community it is there to serve.

(f) Personnel management and career planning are inadequate.

For these and other defects the central management of the Service, the Treasury, must accept its share of responsibility.

2. We propose a simple guiding principle for the future. The Service must continuously review the tasks it is called on to perform; it should then think out what new skills and kinds of men are needed and how these men can be found, trained and deployed.

B. Memorandum no. 5 submitted by the Treasury. Vol. 5 (1), pp. 20–5.

Background – the three classes
2. There are three Scientific Classes which are described in detail in the Introductory Factual Memorandum.* Briefly, the Scientific Officer Class, about 4,000 strong, has the main responsibility for scientific advice, research, design and development in the Civil Service. Its pay has long been aligned with that of the Administrative Class, although the grades of the two classes do not exactly

* Volume 4, Memorandum No. 1.

correspond. Most recruits are first- or second-class honours gradu-ates or holders of postgraduate degrees in scientific subjects. Those with degrees of this standard in engineering are also eligible for appointment and are employed mainly on research and develop-ment work. The Scientific Officer Class is assisted by the Experi-mental Officer Class (about 7,500 strong), which works under its general guidance, assisting in new investigations, particularly in their detailed organisation and execution, and doing practical work requiring the application of established scientific principles. Younger recruits to the Experimental Officer Class need qualifica-tions at GCE 'A' level standard; older ones (except those who are already established Scientific Assistants) have to have a degree, HNC or other equivalent qualification. In recent years 17 per cent of all recruits to the class have had a degree or diploma in tech-nology, and another 26 per cent have had at least HNC or equiva-lent qualification. Finally there is the Scientific Assistant Class (5,700 strong) to which recruits must have qualifications at the GCE 'O' level standard and a year's appropriate experience. Scientific Assistants conduct experiments and tests under the general supervision and instruction of Experimental Officers, whom they also assist generally.

3. The Tennant Committee on the Organisation of the Scientific Civil Service, 1965, considered whether the Scientific Officer and Experimental Officer Classes should be amalgamated, but con-cluded that at that time the balance of advantage lay in retaining the present arrangements. They added that other recommenda-tions in their report would not make it more difficult to unify the two classes if at any time it were considered desirable.

4. We have assumed that, despite this recent study, any Com-mittee looking at the structure of the whole Civil Service would wish to reconsider this question, especially as the Tennant Com-mittee was working against the background of separate Administra-tive and Executive classes. The structures of the Scientific Classes and of the management classes are similar in some respects and are both designed to recruit at different educational levels. Thus one is naturally led to consider whether a single structure corresponding to that proposed for the Administrative, Executive and Clerical Classes could with advantage be adopted in the scientific field.

7. The balance of these arguments seems to us to be against a merger. We approached the problem with an open mind; but we have been led steadily to the view that the Tennant Committee was right and that its conclusion still holds good, despite the fact that some of the considerations have changed since it reported.

C. Report to the Fulton Committee of a Management Consultancy Group.

As opposed to the Treasury submission, the Management Consultancy Group concluded:

166. To sum up, the overall impression that we gained was that the Scientific Officer Class occupied a highly privileged position. To a large extent this was acknowledged and was thought to be essential if the Service were to continue to attract scientists of great ability. It was reinforced by the arrangements for fluid complementing and individual merit promotions which we describe under general comments on the specialist classes at the end of this section. Above all, however, the outsider is surprised to find such a complex hierarchy of classes and grades with each class based on academic qualifications on entry in a field where innovation, often innovation by teams which include outstanding individuals, is the vital product. In this type of activity, cooperation, flexibility in the use of staff and individual motivation are all-important and formal definitions of duties often have little practical meaning. This was in fact borne out by the flexible use of staff and the practice of assigning a place on the team to the man who could do the job rather than by strict adherence to the formal roles of the classes and grades. If anything, the formal definitions of duties tended to militate against the smooth completion of the task in that they led to the difficulties and complexities referred to in paragraphs 161 to 165.

167. These complexities could be avoided by the abolition of the separate classes in the Scientific Civil Service and their replacement by a grading system based solely on job content. The optimum arrangements for grading scientific work require further study and, as we develop later, should, in our view, form part of a wider exercise designed to introduce unified grading arrangements throughout a classless Civil Service.

D. Memorandum No. 152 submitted by Sir Frank Turnbull, formerly Deputy Under-Secretary of State, Department of Education and Science.

Turnbull put forward a different personal point of view from that of the Committee.

1. For a long time I have felt that the fact that the Scientific Civil Service is constituted so closely on the model of the Administrative Civil Service, is a major misfortune. The idea that Scientific Civil Servants should be recruited for permanent service, and be paid on scales broadly comparable with those for other comparable classes derived from the desire to give the Scientific Civil Service 'parity of esteem'.

2. Scientific work is quite different in character from administrative work. Administrative work for the central Government is a highly specialised occupation, and there are great advantages in giving people a permanent career in it. For one thing, they cannot easily flounce out saying that they disagree with Government policy. For another, accumulated experience is of considerable value, and Ministers would have less confidence in the Service if people in it were in the habit of moving in and out frequently. To spend five or ten years in the Administrative Civil Service does not, in general, make a person more eligible for outside employment, and he would not enter it if there were not a firm prospect of a full career.

3. But on the Science side, the same considerations do not, in the main, apply. The bulk of the Scientific Civil Service is recruited to do scientific work, and much of it is research. While there are exceptions, most people do their best research before the age of 45, and some do it while they are quite young. There is then a major problem as to what a good research man will do in later life. Research administration and the scientific administration in Departments cannot absorb them all. It will not do the Administrative Civil Service any good if it is flooded with people from the Scientific Civil Service who have ceased to be good at research, and have no qualifications at all for administrative work. In my experience, comparatively few have any real aptitude for administration as practised in Whitehall, and even those that have labour under considerable handicaps. This points to the need for arrangements which make it possible to employ scientists in government research and subsequently transfer them to teaching and industry.

4. British industry badly needs more people well trained in scientific disciplines. At present, most of the good graduates in science go into the universities or Government-financed research establishments. They get on to a Civil Service type of career with the right to go on to 60, or 65, and, at least, considerable difficulty is placed in the way of moving out to industry, or elsewhere, without personal loss. This is partly because, in the middle grades, the Scientific Civil Service is rather better paid and offers assured increments to a higher level than obtain in comparable posts in industry. As against this, of course, opportunities at the top in industry are much greater. But, for the ordinary competent scientist, Government work offers a more assured, and probably more agreeable, career.

5. The Research Councils are under pressure to conform to the terms of service which prevail in the Civil Service. Only the Medical Research Council has a substantially different system. It pays to scientists rates broadly comparable with the Health Service or universities, according to whether the individual's interests are directly clinical or not. MRC superannuation arrangements allow their scientific classes to move easily into universities or the Health Service and vice versa. The MRC's normal practice is that initial recruitment of scientific staff should be on short contracts. At a suitable age, after not less than five years' service, they determine whether a man is suitable for a lifetime of research, and they keep only a proportion of those they take in on this basis. I believe these arrangements are one of the main reasons for the high reputation of the Medical Research Council. The best people are willing to work there, because they know that they can retain real freedom to move to the outside world. The Science Research Council, the Agricultural Research Council and the Natural Environment Research Council are tied to the Civil Service system. It would be very desirable for the Scientific Civil Service to move in this direction, so that easy transfer could occur to the university, industry and schools.

7 Parliament and Science

One of the constant concerns of scientists, administrators and politicians who have interested themselves in the making of science policy has been the problem of deciding what contribution Parliament can or ought to make to this work. Although in the extracts that follow we give the greatest space to the events of recent years, this concern has a long ancestry. Thus C. M. Douglas writing at the beginning of the century pointed out the great discrepancy between scientific method and Parliamentary 'method':

The mood of Parliament is as remote from science as its procedure is from celerity. Not that, for example, the House of Commons is indifferent to expert opinion. It is a body largely composed of men who are accustomed to rely on trained advisers; and nothing is more characteristic than its willingness to hear people on subjects which they understand. But it is in practice much more a medium for their criticism and accommodation of ideas than for their origination. It is infinitely sensitive to the movements of public opinion, but it almost never initiates anything, and it does not even formulate ideas till they have long been the common property of educated people.*

In the course of this chapter we look at the relations of science and Parliament from three aspects.

Parliament has traditionally attracted few scientists or technologists. Throughout the nineteenth century as well as in the present there have been significant exceptions – men who have promoted scientific interests inside as well as outside the House. In particular, scientists such as George Stokes (a president of the Royal Society), Lyon Playfair, Michael Foster and A. V. Hill come to mind. But at all times such men have been heavily outnumbered by representatives of other professions and occupations; thus an analysis of the professions of Members elected in 1966 showed the following:

scientists, engineers	14
medical profession	8

* In *Science and Public Affairs* (ed. J. F. Hand, G. Allen, London, 1906), pp. 213–14.

lawyers	109
lecturers, teachers	62
managers, economists	27

(It is important to realise, however, that the 'label' attached to a particular Member is not necessarily indicative of either his interests or his degree of activity in the House once he is elected.) Nevertheless, it is legitimate to ask what the individual scientist or engineer in Parliament might hope to achieve [7.1]. Most probably, by himself, not a great deal today, although for the Member with sufficient drive and tenacity there is always scope for what amounts to a one-man campaign.* But the Member is unlikely to want to operate completely alone; he will almost certainly join one or more of his party's backbench subject groups and possibly also one or more cross-bench interest groups. Little is known about these groups since their activities are confidential and they rarely emerge into the public arena; in consequence their work is not reflected in the following extracts. Not surprisingly, the activity of these groups is roughly in proportion to the intensity of interest in issues of science policy in the community as a whole. The best short account of the work of the backbench groups is that given by Vig.†

An important point to be noted about the backbench committees is that they are unofficial; thus they lack the considerable powers customarily accorded to Parliamentary Committees (see below). Also unofficial in this sense is the Parliamentary and Scientific Committee, which constitutes the second feature of this chapter. The Parliamentary and Scientific Committee is a legacy from the 1930s when the 'unscientific' nature of politics aroused concern among scientists increasingly anxious about the social implications of their work. Throughout the 1920s the British Science Guild and the Association of Scientific Workers [7.2.1] campaigned through their Parliamentary committees for a scientific committee within Parliament to act as a watchdog. It was not until 1933 that, with the support of such bodies as the Institute of Physics, a liaison committee was formed with the title of the Parliamentary Science Committee. The same year, the chairman of the British Science Guild, Sir Arnold Wilson, was elected to Parliament and became the first chairman of the new Committee, gathering around him interested Members from each party [cf. 4.4.3]. At the outbreak of the second world war the Committee decided to disband, a decision which

---

* The case of Joseph Hume and the estimates of public expenditure in the first half of the last century is frequently held up as a case of what self-help can achieve.

† N. J. Vig, *Science and Technology in British Politics* (Pergamon Press, Oxford, 1968), pp. 115–20.

appears not to have been unanimous so that a number of the dissidents formed a new body in November 1939 with the title the Parliamentary and Scientific Committee. The role of the Committee has been well set out by a former chairman [7.2.2]; its formally stated aims and objectives are:

(*a*) to provide Members of Parliament with authoritative scientific information from time to time in connexion with debates.

(*b*) To bring to the notice of Members of Parliament and Government Departments the results of scientific research and technological development which bear upon questions of current public interest.

(*c*) To arrange for suitable action through Parliamentary channels whenever necessary to ensure that proper regard is had for the scientific point of view.

(*d*) To examine all legislation likely to affect the above and take such action as may be suitable.

(*e*) To watch the financing of scientific and technological research, education and development.

(*f*) To provide its members and other approved subscribers with a regular summary of scientific matters dealt with in Parliament.

From time to time the Committee has undertaken certain 'campaigns' in accordance with these aims and the second world war was a particularly active time for it in this respect. As an illustration of one of these we may take its efforts to ensure that the most effective use of scientific manpower was made by the government during the war [7.2.3]. But as to the Committee's overall effectiveness one commentator at least has taken a rather sceptical view [7.2.4].

The third aspect of the confrontation between science and Parliament that we shall consider is the device of the Parliamentary Select Committee. These are established by decision of the whole House and are customarily delegated with the power 'to send for persons, papers and records'. A Select Committee exists in order to make inquiries; it reports its findings back to the House but cannot compel the House to take action on them. Thus such a Committee has a primarily investigatory and clarificatory role. There are many precedents for the use of this device to examine subjects of a scientific or technological nature, for instance weights and measures, explosions in steamboats, the gas industry and mining accidents.* These inquiries have usually been of a 'one-off' nature, that

* Austen Albu, in *Minerva*, Autumn 1963, vol. 2, no. 1, pp. 1–20 gives many more examples.

is, the Committee having investigated and reported on its allotted subject it would not normally be reappointed. Other Select Committees of the House, however, are reappointed sessionally, for example the Committee of Public Accounts (which dates from 1861), the Estimates Committee (1912) – now superseded by the Select Committee on Expenditure (see below) – and the Select Committee on the Nationalised Industries (1957). The first two of these are (or were) concerned mainly with finance; the Public Accounts Committee looks at the use made of money voted by the Commons and decides whether it has been properly spent, and the Estimates Committee examined proposed forms of expenditure with a view to deciding whether they permitted economies. Insofar as policy-making is inseparable from financial control it is apparent that the attempt to separate financial inquiries from those involving issues of policy will be difficult, and so it has proved in practice. Thus when these two Committees have looked from time to time at departments or functions involving science and technology their reports have a bearing on science-policy formulation [7.3.1].

In the past decade, criticism of Parliament has greatly increased. The burden of it has been that, as an institution, Parliament has become more and more ineffectual when pitted against the executive, and that as regards science and technology it has lacked both the expertise and the access to information which alone would enable it to form a considered view on issues which are coming to affect society more and more. To redress this imbalance between Parliament and government the device of a Select Committee to cover science and technology has been widely suggested as a solution, albeit not perfect in itself [7.3.2]. Other Parliamentarians, reflecting their interest in science and technology as factors in economic growth, proposed a broader field of inquiry [7.3.3]. More specific proposals along these lines were offered to the House's Select Committee on Procedure in the session 1964–5 when it was examining ways of reforming the Estimates Committee [7.3.4, 7.3.5]; the proposals were not accepted although the Committee agreed that the machinery of Parliament needed to be improved to enable Members 'more effectively to influence, advise, scrutinise and criticise', therefore a new Select Committee should be set up as a development of the Estimates Committee to examine 'how the departments of state [considered individually] carry out their responsibilities'.

In session 1965–6 a measure of specialisation, along the lines recommended by the Select Committee on Procedure, was introduced by the Estimates Committee when seven Subcommittees were set up. Among these were defence and overseas affairs, economic affairs, and technological and scientific affairs. The same Subcommittee structure was maintained in the following session but it was abandoned in session 1967–8 (and the system of letter titles – Subcommittee A, B, etc. – reintroduced) because the overall strength of the Estimates Committee was reduced and by then the new specialist Select Committees (see below) had been formed and it was thought desirable not to give the impression of needless duplication. One report came from the Subcommittee on Technological and Scientific Affairs during its relatively short life.*

In the midst of this activity, an outside commentator was advocating a more radical solution on the grounds that a non-partisan approach to science [cf. 7.2.1] would effectively kill its significant discussion [7.3.6]. There is little, if any, evidence, that many MPs saw Parliamentary reform in quite these terms, but ample that the suggestion to set up more Select Committees of an investigatory character was widely supported.†

The object for which these Parliamentary reformers were striving was finally realised in 1966 when a Select Committee on Science and Technology was announced [7.3.7]. For the first time in Britain scientific issues would be subject, through the presentation of evidence from government and the Civil Service, to detailed and continued scrutiny by Members of Parliament [7.3.8]. The experiment has, it is widely agreed, been successful, in that the inner workings of some departments have been cautiously revealed [7.3.9].

The Select Committee's justification was called into question by another report from the Select Committee on Procedure in 1969 which proposed a reformed Estimates Committee under the title of the Select Committee on Expenditure with a remit to examine departmental efficiency. The Science and Technology Committee was strongly defended by one of its members in the subsequent debate on the report [7.3.10]. The Conservative government returned in 1970 accepted the substance of the Procedure Committee's proposals but decided in favour of retaining the Select Committee on Science and Technology (among others) for the life of the current Parliament at least.‡

---

* 13th Report from the Estimates Committee: Space Research and Development, House of Commons Paper 601 of 1966–7.
† See, for instance, the survey results presented by A. Barker and M. Rush in *The Member of Parliament and His Information* (Allen & Unwin, London, 1970, for PEP and the Study of Parliament Group).
‡ Select Committees of the House of Commons, Cmnd. 4507, October 1970.

7.1 SCIENCE AND THE INDIVIDUAL MEMBER OF PARLIAMENT

7.1 The Engineer in Parliament

From Arthur Palmer, MP, *Electronics & Power*, April 1969, pp. 120–2.

The author is an electrical engineer, a former chairman of the Parliamentary and Scientific Committee and was the first chairman of the House of Commons Select Committee on Science and Technology [7.3.7, 7.3.9]. In this article he discussed the traditional attitudes of engineers towards politics.

The professional engineer on the whole shows diffidence and caution in regard to politics. Resentfully he seems to accept that policy decisions in relation to Government are necessarily taken by that ill-defined force 'they'. He does not see himself as part of the political process as do many other professional groups. Professional engineers are relatively rare birds in Parliament. They are not as numerous as, say, economists, lawyers or teachers.

What is the reason? Is it temperament? I think that in part it is. Technical people are perhaps not natural talkers. They find their satisfactions in hardware – in designs, quantities and mathematical affirmations. They distrust the orator and find people in the mass and their problems troublesome. They dislike the expression of ideas and principles not matched by some corresponding action. I suspect that the complexities, emotionalism and subtleties of politics repel many technical minds.

Then again, is the apparent reticence of engineers in relation to politics based on a misunderstanding about the nature of politics? It may be. I think engineers are prone to equate public administration with politics, which is wrong. Politics is really about conflict in human society, and the politician is concerned with the resolving of that conflict as peacefully as he can. Because he is concerned with conflict, the politician has something in common with the soldier. Note the wealth of military metaphors in political speeches and exhortations. Your engineer does not always see that political parties have their history, traditions, interests and philosophies, all of which affect their actions as much sometimes as do the administrative needs of a situation.

Political parties are armies of a kind and not simply suppliers of alternative boards of directors to run a country instead of a company. Your engineer may therefore easily repudiate politics, because politics cannot, in the nature of things, conform to his known rational world, where objectivity is king.

A third possible (and very practical) explanation for the caution of engineers about politics is that their career structure makes it hard for them to embark on the uncertainties of an MP's life. The processes of democracy being as they are, at every General Election a large proportion of the membership of the previous House of Commons is declared redundant by the voters.

A final reason, I think, for the weakness of the engineer in relation to public life has been the past shyness of the major engineering institutions, which speak for his professional interests (or should do so). When Governments and Parliaments have taken action in fields of obvious professional concern to engineers, the institutions have often remained silent.

Neither medicine, law nor teaching has shown the same degree of genteel modesty. These other professions have known how to go down into the market place and make themselves heard in the busy world of men and affairs; engineers unfortunately have not had the flair. But, to be fair, this statement is not so true as it once was. The useful meetings that take place these days between technically interested parliamentarians and leaders of the chartered institutions are one indication of a changing situation; the occasional appearance of articles such as this one in institution journals is perhaps another indication.

7.2 THE PARLIAMENTARY AND SCIENTIFIC COMMITTEE

7.2.1 Science is not Partisan

From H. W. J. Stone, 'Science in Parliament', *Progress*, July–September 1933, pp. 214–16.

H. W. J. Stone was the Parliamentary Secretary of the Association of Scientific Workers and was responsible for agitating during the early 1930s on behalf of the Association for more effective communication between scientists and Parliament. In a persuasive article he described the non-partisan nature of science.

That there ought to be some machinery for the purpose of establishing perfect liaison between Parliament and Science is a proposition which few would gainsay; indeed, it is axiomatic. It is an ideal to which much lip-service has been paid, but little more than that. Much might be written to demonstrate the desirability of such liaison. The effort would only amount to pushing an open door; the outlook for achievement would be far more promising if that door were *not* open. To have aspirations accepted, without their being translated into action, is a far less healthy situation than to encounter a virile opposition which engenders warfare. After warfare there is necessarily a peace conference when there has to be mutual give and take. An unchallenged ideal which is not translated into concrete action only finds an inglorious resting place in 'no-man's-land'.

Before Science can establish effective liaison with the Parliamentarians a tremendous gulf will have to be bridged. What is that gulf? To that question the answer is cynical, but none the less true. The politicians are not yet convinced that Science controls enough votes to make it matter. Votes dominate the atmosphere in which politicians regard everything. They are the politicians' oxygen. Roughly speaking, the electorate of the United Kingdom is thirty millions. At a liberal estimate the scientists of the United Kingdom number thirty thousand – or one-thousandth part of the whole! It has been said that under a democracy heads are counted rather than what is in them; and Augustine Birrell has remarked: 'Minorities must suffer: 'tis the badge of their tribe.' Nevertheless the sufferings of minorities frequently prove to be their salvation. Sufferings beget clamour: clamour begets attention: and attention precedes achievement. But the clamorous must know what they want, and must speak with one voice. Having mastered that fact the world is your oyster – whether you be a majority or a minority. We have all watched the methods of the 'stunt' Press, getting their readers to sign forms of demand and forward them to MPs. Not infrequently Members yield to the clamour without staying to consider that the requests come from a minority. They mistake clamour for *voces populi*, and fear loss of votes. Some years ago there was an agitation of this character, and a Member asked his secretary how many communications had been received. He was told 300 – roughly 200 for, and 100 against. He said at once: 'If I wish to retain my seat I must support this measure.' His secretary remarked that that conclusion was premature, pointing out that of a total electorate of 40,000 only 300 had thought the matter vital enough for a letter either way; 39,700 didn't care a hoot about the question one way or the other – or, at any rate, had given no sign. So much for being swayed by clamour!

Attempts have been made from time to time to form a Science Committee in Parliament, just as there is a Medical Committee, an Agricultural Committee, a Commercial Committee, and so on. Members have been approached to join and have consented. On paper the Committee has looked imposing. When, however, meetings of the Committee have been summoned, the attendance has been distinctly disappointing. Why? Members do not care to refuse to join such a Committee. It might alienate votes. On the other hand, having no real interest in Science, they find the usual excuses for absenting themselves from meetings of the Committee. There are no division lists published to reveal their attendance or non-attendance. If the work of the Committee is referred to in their constituencies they affect a sympathetic interest, and say 'Oh yes! I am a member of that Committee'. That is the sort of Committee that travels nowhere. In the case of the Agricultural Committee or the Commercial Committee some decent show of participation is enforced because so many votes depend on a Member's attitude thereto. In Science it is otherwise. One vote in a thousand is hardly likely to make the difference between re-election and rejection.

To break down this inertia Science must storm the portals of Parliament *en masse*, having a clearly defined programme, speaking with one voice, and aided by a propaganda which will carry the bulk of the thinking electorate in favour of its proposals. Such a result is not impossible of achievement. The various scientific organisations throughout the country should convene a conference in London consisting of duly accredited delegates from each organisation. At that conference the broad outlines of a national scientific programme should be discussed, leaving the sectional interests to hammer out their respective contributions to the programme by subcommittees appointed thereat. It does not much matter which organisation acts as Convener of this conference. It might be the British Association; it might be the Royal Society; or it might be the Association of Scientific Workers. The great thing is to get it called, and to find a suitable Chairman – a statesman as well as a scientist. Why not Mr H. G. Wells? He has vision, and possesses the happy knack of getting things done. Let plenty of time be taken; present ultimately a workmanlike programme, and no Member of Parliament will – with an eye to votes – be able to afford to stay outside the movement. Not only that; but he will perforce have to attend the meetings of the Committee when formed. A real Parliamentary Science Committee could achieve wonders, and both Houses would hasten to perform its behests. To know what one wants is half-way to accomplishment.

What can be achieved in the Parliamentary field when scientific and commercial bodies join hands to exert pressure on Government Departments has been demonstrated during the current session in connexion with the renewal of the Cotton Industry Act. The third reading of this Bill in the House of Commons on 19 May was, with one exception, a prolonged paean of praise for the accomplishments of research in the cotton industry. The Government spokesman (Dr Burgin, Parliamentary Secretary of the Board of Trade) was almost fulsome in the bouquets he cast at the feet of the Goddess of research, and waxed lyrical in describing the benefits to the cotton industry. The third reading was moved by a Socialist on behalf of a Conservative Member. The Socialist Leader gave a reasoned support, and the third reading was passed without a division. Why? No Lancashire Member dare oppose the Bill. More than that, no Lancashire Member dare refrain from supporting the Bill. Votes depended on it. It was one of those inconvenient measures which cut right across political party divisions. It was a bread and butter question. So is Science!

7.2.2 The Role of the Parliamentary and Scientific Committee

From M. Philips Price, 'The Parliamentary and Scientific Committee of Great Britain', *Impact of Science on Society*, Winter 1952, vol. 3, no. 4, pp. 258–9.

M. Philips Price, a Cambridge science graduate, had a varied career as a journalist before entering Parliament. He was a Member of Parliament from 1929 to 1931 and was re-elected in 1935; he was chairman of the Committee from 1946–52.

It is always difficult for people in one country to understand exactly how the parliament and constitution of other countries work in practice, and the British Parliament and the unwritten constitution of Great Britain present particular difficulties in this connexion.

One starts off with the fundamental difficulty that the committee system which prevails in the great majority of European and the American parliaments does not exist in the British Parliament at all. Whereas in the USA and in most European parliaments there are important functions attaching to such committees as the foreign affairs committee, finance committee and so forth, the British Parliament provides no such parallel.

There has always been a fundamental reluctance in Britain to allow committees of Parliament to exercise power over the government of the day and over the executive. This power belongs, as it is conceived, to the full assembly of Parliament alone. It is, of course, often necessary, owing to the pressure of business, to delegate some of the work of Parliament to committees. Bills are frequently sent to what are known as standing committees which are composed of members of all parties in the same proportion as in the whole House. But standing committees have to report back to the whole House, which may alter what the committee has done. On the other hand the budget is never sent to a special committee, nor are the Foreign Office estimates nor the estimates of the great supply services. These are always dealt with by the whole House, which resolves itself into a committee of the whole House for the purpose. . . .

For the rest, committees in the British Parliament mainly consist of comparatively informal groups, either of members of one party or of all parties, concerning themselves with particular subjects. These are virtually study groups and merely help to provide a focus for discussing and forming views about particular issues. . . .

The Parliamentary and Scientific Committee is one of these informal all-party groups, but it has, for one reason or another, managed to establish itself with more continuity and formality. It is however, very definitely an unofficial body and an all-party one, membership of which is open to members of any party in both the House of Lords and the House of Commons, and also to the nominated representatives of those scientific and technological organisations in the country which, under its constitution, can be affiliated. No organisation can be a member of it which is engaged in promoting some scientific process or project for profit or commercial gain. Societies must be engaged in research or scientific education or be concerned with protecting or advancing the interests or professional well-being of those engaged in scientific work.

The Parliamentary and Scientific Committee receives no subvention whatsoever from the government or from any political party, and its modest revenue is drawn from the voluntary subscriptions of the peers and members of Parliament who belong to it and also from the subscriptions of the affiliated scientific bodies. Its approximate annual revenue from these sources does not broadly amount to more than £1,500 a year, a sum which is, of course, only sufficient to enable it to maintain a part-time staff and to organise a limited number of meetings, deputations, etc. and generally to stimulate a certain amount of useful activity in the Parliamentary field so far as scientific and technical issues are concerned.

The committee has sought to become a centre in the British Parliament for the consideration of any scientific or technological question which may possibly have a useful bearing on the current or future activities of Parliament or government. Those who founded the committee thought that very definite benefits might result if public-spirited representatives of the various societies concerning themselves with science and technology could meet together from time to time with a voluntary group of members of Parliament and peers who appreciated the important contribution which science can make to the government of any country in modern times, so that both Parliament and government departments should receive a continual stimulus and encouragement to take into account what science and technology can contribute to a wide range of public affairs.

7.2.3 The Parliamentary and Scientific Committee and the Utilisation of Scientists in Wartime

From M. Philips Price, 'The Parliamentary and Scientific Committee of Great Britain', *Impact of Science on Society*, Winter 1952, vol. 3, no. 4, pp. 269–71.

An illustration of the type of 'campaign' the Committee has from time to time undertaken.

This was a matter of constant concern to the Parliamentary and Scientific Committee during the period 1941–3 and quite early in the war members had serious misgivings about the failure to make proper or coordinated use of the scientific and technological resources of the country.

The matter was first raised by Viscount Samuel, a very active member of the committee, in the House of Lords on 2 April 1941, when he rose to ask the government 'whether they can make any statement on the extent to which the assistance of scientists has been enlisted in the prosecution of the war; and to move for papers'.

He urged that greater use should be made of industrial men of science such as chemists and physicists engaged in industry, or in practice as consultants. He referred to the criticism that scientists in government departments were not sufficiently in touch with operations, and asked what steps were being taken to secure co-operation between American and British science in the prosecution of the war.

There was also considerable correspondence during 1941 between the chairman of the committee and the Lord President of the Council on this subject, and a subcommittee was set up to draft detailed recommendations. Gradually things began to improve, and at the committee's annual luncheon on 3 February 1942, the distinguished scientist, Sir Henry Tizard, pointed out that although there was this large amount of scientific talent available to the government, he doubted whether the right use was being made of it in all cases. He said:

So far as the strategy of science is concerned, I am not so confident. Much has been done to improve it, but much remains to be done. The strategy of science is again similar to that of war. The strategy of pure science is to attack at the weakest spot of the barrier to knowledge. The secret of science is to ask the right question, and it is the choice of the problem more than anything else that marks the man of genius in the scientific world. The strategy of science applied to war is to attack at the point where the dividends are greatest in relation to the effort. Our tactical strength is great, but it is not unlimited. We cannot afford to dissipate our efforts over things that do not matter or do not matter much, and we must remember, too, that any technical advance, to have a decisive effect in war, makes big demands on the productive capacity of the country.

He accordingly urged the Parliamentary and Scientific Committee to concentrate on ensuring more effective cooperation between all branches of the government and its scientific advisers.

After the subcommittee had evolved detailed recommendations on the issue, a deputation was appointed to wait on Mr R. A. Butler, MP, who had recently been appointed chairman of the government Scientific Advisory Council, and whom it saw on 16 July 1942. The main point urged by it was the setting up of a more effective scientific organisation to inspire the scientific and technical direction of the war. Mr Butler undertook to pass on the views expressed to the Lord President of the Council and the Minister of Production.

Further action was taken by the parliamentary members of the committee, by the placing of a motion on the Order Paper of the House of Commons on 28 July 1942, stating that the effective prosecution of the war required the early establishment of a central organisation to coordinate research and developments in relation to the war effort and to ensure that the experience, knowledge and creative genius of British technicians and scientists exert a more effective influence over the conduct of a highly mechanised war. That motion received the support of 145 members of Parliament of all parties.

The main outcome of all these representations was undoubtedly to cause a considerable improvement in the government's arrangements for the scientific direction of the war and one of the most important of these was the decision to appoint three leading scientists as scientific advisers to the Minister of Production.

With regard to the appointment of these advisers the Parliamentary and Scientific Committee unanimously passed the following resolution at its meeting on 8 September 1942:

This committee, while welcoming the appointment of three full-time scientific advisers to the staff of the Ministry of Production insofar as it establishes the nucleus of a central scientific and technical board, regrets that their field of activity is apparently to be limited to the sphere of production and does not include the scientific and technical activities of the service departments or the other ministries outside the strict field of production. An extension of its functions is needed to ensure that all scientific considerations are coordinated and given full weight over the whole field of the national effort. The committee considers, therefore, that in order to cover this wider field, scientific advisers should have direct access to the War Cabinet and that accordingly the Lord Privy Seal should exercise his supervisory functions over the new body directly on behalf of the War Cabinet.

This was sent to the appropriate ministers and the press, and Sir Stafford Cripps, Lord Privy Seal, replied as follows:

I think that it would be wise at the present time not to press for any more precise definition of the functions of the scientific advisers who have been appointed, but rather to let me handle the position as best I can with a view to their gradual establishment and the extension of the scope and value of their work. Very often if one tries to make the directives on this sort of question precise, one only ends by limiting rather than extending the powers.

7.2.4 Verdict on the Parliamentary and Scientific Committee

From S. A. Walkland, 'Science and Parliament: the Origins and Influence of the Parliamentary and Scientific Committee II', *Parliamentary Affairs*, 1963–4, vol. 17, pp. 399–400.

Some of the Parliamentary and Scientific Committee's more recent work – on the financing of Research Associations, for example – is still at an inconclusive stage and has therefore not been dealt with. Enough has, however, been said for a general picture of the Committee to emerge. Over the last quarter of a century it has followed a familiar British path. From a fairly radical Parliamentary innovation, sponsored by Left-wing scientists, it has become an 'institution', with distinguished patrons, an honorary membership, annual luncheons at the Savoy and a page to itself in the HMSO publication, *Britain – an Official Handbook*. As such, it is perhaps understandable if somewhat inflated claims are occasionally made on its behalf. The recent editorial in *Nature* on 'Science and Parliamentary Efficiency',* is an example of some general misunderstanding of the Committee's modest functions and influence. It claimed that the composition of the Committee 'is both a counterweight to party political views on any particular issue and offers some safeguard

* *Nature*, 27 April 1963, vol. 198, p. 321.

against reliance on a single scientific point of view'. But as should be apparent from the preceding pages, these archetypical fears of scientists when faced with political processes cannot be allayed by the Parliamentary and Scientific Committee. It has never tackled matters which are the subject of fundamental dispute between the parties, and it lacks both the status and organisation to make detailed technical assessments of particular policy decisions even where there has been suspicion that the political element in the decision has been allowed to outweigh technical considerations. Zeta, much of the civil nuclear programme, marine nuclear propulsion and the transonic civil airliner are projects which might have repaid more informed Parliamentary scrutiny. The Committee is not a research body, however, although it is capable of fact-finding in some fields, but it is worth noting that even in the sphere of technical education, where the Committee can claim most success, its policy relied more on informed 'hunches' than on applied educational research. To adopt the terminology often used of its executive counterpart, it is a committee for science, and not a scientific committee.

As a defence organisation for scientific interests it has had, and still has, an important role; as a detached Parliamentary body surveying the whole field of government endeavour in scientific research and development it has been in a position to suggest priorities and to draw official attention to what the Advisory Council on Scientific Policy has called the 'fallow fields'. It has undoubtedly played a part over the last two decades in changing the focus of attention of Parliament, but the increasing public and political awareness of science is depriving the Committee of its role as a general lobby for science in government. Instances of the way in which this is happening have been given; a very recent significant pointer is that the Institution of Professional Civil Servants, almost a founder member of the Parliamentary and Scientific Committee, has carried its case against the Trend Report on the organisation of civil science to the Labour Party rather than to the Committee. Given the elevation of science in political status this is a natural course, but one with severe implications for the future of the Committee.

The Parliamentary and Scientific Committee is perhaps best seen as a well-meaning, but essentially amateur, reaction, of a rather familiar and depressing type, to a particular historical situation. Given the rudimentary nature of the official organisation for science in the 1930s and 1940s, the Committee was an adequate means of bringing Parliamentary pressure to bear on the executive in the field of scientific policy. But as the direction of scientific research, development and education becomes more integrated, and the departmental structure more monolithic, the scope for any unofficial Parliamentary agency to exert influence on policy is likely to be restricted to vanishing point.

7.3 SELECT COMMITTEES

7.3.1 Parliamentary Finance Committees and Science-Policy Making

From N. J. Vig and S. A. Walkland, 'Science Policy, Science Administration and Parliamentary Reform', *Parliamentary Affairs*, 1965-6, vol. 19, pp. 289-90.

In this article, published before the specialist Select Committees were set up at the end of 1966 [7.3.7] but after the Estimates Committee was allowed to set up 'departmental' Subcommittees [7.3.4], Vig and Walkland point out the defects of the Estimates and the Public Accounts Committees as instruments to exercise a watching brief over science and technology and go to suggest an improvement in the current situation.

It would be useful at this stage to develop a distinction which is seldom insisted on in this debate, between, on the one hand, the traditional bipartisan concerns of a legislative assembly, centring mainly around the financing of government and its associated interest in administrative efficiency and 'value for money', and, on the other, concern over larger questions of national science policies and their social and economic implications, of a type which may entail political dispute. It is to the former that the Government has tried to deflect Parliamentary attention, by endorsing specialisation of the Estimates Committee rather than the formation of a select science committee. It would be difficult, however, to sustain that either the Estimates or the Public Accounts Committees has been successful in investigating R & D activities and science agencies. The Estimates Committee has made a dozen or so reports in this field, mainly on civil science, but these were narrowly drawn, and their contribution to the special administrative problems of science agencies has been insignificant. The Public Accounts Committee is rigidly restricted to details of contracts and financial management, and its reports on the financing of R & D activities have been particularly narrow in scope. Neither committee has developed either standards or techniques for the assessment of the effectiveness of science programmes, based on cost/benefit criteria or other general considerations apart from administrative and financial accountability. Although both the costing of R & D work and the assessment of probable benefit are complex and essentially speculative enterprises, entailing analytical and evaluation techniques which differ considerably from normal administrative practices, and requiring departmental budgeting on a more sophisticated pattern than normal government accounting, the criteria can be generalised, as can the procedures and criteria which are applicable to the allocation of resources between conflicting projects at the level of the Department or agency. It could presumably be a function of a specialised Estimates subcommittee to attempt to develop conscious appreciation and application of such procedures and criteria. At the level of the national government, a specialised Estimates Committee could similarly, and without strain to its order of reference, help to specify systematic procedures, such as a science budget, financial and manpower inventories, machinery for the coordination and control of inter-Departmental programmes, manpower planning, which would allow the executive to recognise earlier and more clearly omissions, wastages and imbalance in the national scientific effort and to take the necessary remedial steps.

7.3.2 The Member of Parliament, the Executive and Scientific Policy

From Austen Albu, MP, *Minerva*, Autumn 1963, vol. 2, no. 1, pp. 17–20.

The author wrote this paper when he was chairman of the Parliamentary and Scientific Committee. His article was a key statement of the problems confronting Parliament when it attempts to grapple with science-policy issues: it is inevitably placed at a disadvantage compared with the executive and it must itself create the means for reform.

Whatever steps are taken to ensure that Members of Parliament are not scientifically illiterate, they will need greatly improved resources of research and information to do their jobs properly. Although there are still Members who claim themselves satisfied with the present library service and who cling to the gentlemanly view of the Member of Parliament as a sagacious amateur, there is a growing demand, especially among the younger Members, for a properly equipped research and information service, staffed with men and women educated in a broad range of subjects, including the scientific and technological.

Any comparison of proposals for the greater use of committees in the British Parliament with the procedure in the United States Congress must be finally vitiated by the completely differing constitutional systems. Congress is a constituent part of government; Parliament is not. Congress initiates legislation which in Britain is almost entirely the function of government; the expert preparation of Bills is no longer a parliamentary responsibility. By tradition Parliament votes the Supply asked for by government, without detailed consideration of the Estimates, provided the policy which they are to support is approved. None of these considerations would prevent Parliament using the device of the Select Committee to inform itself on scientific matters either generally or in special cases and there is no doubt, from the experience of the Estimates and Nationalised Industries Committees, that a committee of this sort would work in a non-partisan spirit and produce valuable reports. What prevents the development of a committee system, even with these limited objectives, is the degree to which even non-political decisions have become matters of high policy which are taken by the Prime Minister or one of a small number of Cabinet colleagues. The growth of party discipline during the last century has given to the Cabinet, and within the Cabinet to the Prime Minister, a practically unshakeable power in Parliament between elections. Parliament, by allowing this development to take place, has given up the power, in face of the opposition of the government majority, to decide for itself what committees to set up. It is true that once a committee has established itself by tradition over a number of years it would be extremely difficult for a government to oppose its reappointment; but unless a government were willing to share some of its responsibilities for decision making in scientific matters with Parliament, or at least to allow Parliament to learn from the government's own advisers the opinions on which its decisions were based, it would not make a move to set up any new committees for these purposes.

Whether or not the American system has led to a better method for taking decisions on scientific or technological matters by government or for a better control by the legislature of scientific policy, there is nothing to prevent Congress, if it were so minded, from improving its own arrangements. By contrast, if Parliament is to get powers, even to improve its means of obtaining information, there will need to be a softening of party discipline and an acceptance of the view that, on the growing number of matters which in no way involve party opinion, the confidence of the government should not be involved. Informed criticism and even amendment of technical decisions or legislation should not be considered as defeats requiring the government's resignation. Without such possibilities of affecting decisions, the present apathy of many Members of Parliament in face of complex issues will not be overcome.

Cabinet government has great advantages over the congressional system in the task of coordinating policy and administration and nowhere more so than in the complex and often urgent matters of economic, defence and foreign policy. The establishment of a number of Select Committees, with the sole object of obtaining information on the background for policy, would in no way interfere with Cabinet responsibility for political decisions. The growing danger is that the scientist, working behind the shield of the executive, will imperceptibly take over from the electors and their representatives the power of making choices between policies based on alternative values or interests, simply because those policies involve, in one way or another, scientific knowledge. Even governments themselves will find themselves more and more at the mercy of their scientific experts if they prevent the scientific information and advice at their disposal from being exposed to parliamentary and public criticism. The report of a Select Committee does not only inform Parliament; it informs the wider public outside. In this, it performs one of the most vital duties of parliamentary question and debate, without which a democratic system of government will hardly survive.

The reform of parliamentary procedure lies in the last resort, in the hands of individual Members. If a sufficient number of them were to break out of the vicious circle of a growth in executive power, a tradition of withholding information from Parliament, the lack of research services for Members of Parliament and apathy of the latter in the face of the resultant parliamentary situation, no government would be able to resist.

The establishment of a number of Select Committees, charged with the duty of informing Parliament on the scientific component in government decisions, even if supported by greatly improved research and information resources for members, would not, however, as the American experience has shown, remove altogether the growing danger of dominion by scientific advisers and civil servants. To this danger there is no immediate and complete answer. Over time, however, the growing pervasiveness of science in our life and culture should be self-correcting. As more and more people receive at least a partially scientific education, and if the machinery of Parliament is made more vigorous and efficient, it is likely that the composition of Parliament itself will change in the direction of a larger proportion of Members with some scientific or technological background. In this way the views of the experts might become subject to more informed criticism and their power subject to more consistent control. The very process which has engendered the present difficulty might contribute to its cure. It will do so, however, only if some Members of Parliament, supported by enlightened opinion, take an effective initiative in freeing Members from their present domination by the executive.

7.3.3 Standing Economic Committees

From Conservative Political Centre, *Change or Decay: Parliament and Government in Our Industrial Society*, January 1963, pp. 11-12.

This Conservative Party pamphlet appeared before the article by Austen Albu, MP [7.3.2], but insofar as it relates to Parliamentary reform it is not greatly dissimilar in spirit in urging a strengthening of Parliament's investigatory role. It reflects the thinking of those Conservative Members (e.g. Airey Neave – the chairman of the drafting committee – and John Osborn) most involved with issues of science and technology in the period preceding the 1964 general election.

In no sphere is the need for the reform of Parliamentary procedure more urgent than in that of economic policy. This need was already recognised more than a generation ago and brilliantly expounded by Sir Winston Churchill in his Romanes lecture at Oxford in 1930. Sir Winston advocated that Parliament should choose, in proportion to its party groupings, what he called an Economic sub-Parliament, composed of persons of high technical and business qualifications, who would debate the economic problems of the day. Discussion of these usually requires administrative, commercial and technical knowledge and specialised experience. Procedure in this field should now be drastically revised, and it is necessary to carry the argument in favour of delegation to committees a stage further. Is there any reason why a new form of Standing Committee on economic matters should not be formed, which, while responsible to Parliament, could include experts from outside to be called in as advisers?

The House of Commons already has Select Committees on Estimates, Public Accounts, and Nationalised Industries, to examine expenditure in the public sector. It also has study groups for roads and space research which are all-party in character. If the main issues are not essentially party-political, then the growth of such committees is to be encouraged. Committees to examine the main spheres of economic activity would only work, however, if it were possible to avoid the intrusion of party into consideration of such subjects as public investment and incomes policy. This might well prove difficult, though technical and scientific questions arising from unopposed Government legislation could be discussed on this basis. Standing Economic Committees could, however, play a very useful part if the parties were to select back-bench Members with the best industrial or technical experience to serve on them and co-opt industrialists, economists, scientists and other experts to advise them. These co-options might well include members of the new National Economic Development Council. It might also be found desirable, for the cross-fertilisation of ideas, that Members of Parliament drawn from its Standing Economic Committees should serve on NEDC and its subcommittees. It is inevitable that the NEDC should exercise considerable influence on the economic life of the country, and it is therefore important that its work should be not merely an adjunct to Government but properly related to the responsibilities of Parliament to whom it should report.

It is worth considering whether a Standing Economic Committee could not report to Parliament on the technical and other detail of an economic Bill, before its second reading. It may be argued that Parliament, through the Government, has the advice of innumerable commissions and committees on economic and financial matters. The purpose of this suggestion is that, in many economic fields, the two forms of committee – Parliamentary and Advisory – could be combined and the life of the back-bench Member with something to contribute rendered more fruitful than waiting around to support his party in the lobby. Legislation could be more thoroughly considered in advance, and Parliament and industry would thereby come closer together.

7.3.4 Specialist Committees

From Study of Parliament Group, Memorandum Submitted to the Select Committee on Procedure, Appendix 2 to the Fourth Report from the Select Committee on Procedure, House of Commons Paper 303, 1964-5.

The Study of Parliament Group comprises officers of both Houses of Parliament and academics. The extract below comes from a substantial paper suggesting a number of means of Parliamentary reform. In the event the Procedure Committee did not accept the Group's proposal (nor that of the Parliamentary and Scientific Committee [7.3.5]) but recommended (Report, para. 16):

(i) That a new Select Committee be set up, as a development of the present Estimates Committee, 'to examine how the departments of state carry out their responsibilities and to consider their Estimates of Expenditure and Reports.'

(ii) That the new Committee should function through Sub-committees specialising in the various spheres of governmental activity.

21. The Estimates Committee already has terms of reference wide enough to embrace all aspects of the efficient conduct of administration, as evidenced by examples of its Reports during the Session 1963-4. These included in addition to Supplementary Estimates

and Variation in Estimates, topics such as Transport Aircraft, Form of the Estimates of the Defence Department, Treasury Control of Establishments, the Forestry Commission, Services Colleges, Military Expenditure Overseas and the Department of Technical Cooperation. Special Reports relating to Departmental Observations included those on the Home Office, the Administration of the Local Employment Act, 1960 and the Ordnance Survey.

In order to extend the scope of this work we recommend:

(a) that the Estimates Committee should be enlarged to make possible the use of more subcommittees, which would specialise in particular areas so as to cover the whole field of Government vote-borne operations and report on the conduct of administration and on matters necessary for the understanding of policy questions.

(b) Specialist Committees are needed to scrutinise the actions of government in their own fields, to collect, discuss, and report evidence relevant to proceedings in Parliament, whether legislative or other. The main weakness in Parliament's present methods of scrutinising administration, and indeed of debating policy matters, is the limited ability to obtain the background facts and understanding essential for any detailed criticism of administration or any informed discussion of policy. Specialist committees, working on lines similar to those of the Estimates Committee or Nationalised Industries Committee (itself a fairly recently established specialist committee) could go a long way to remedy this. They would be mainly concerned with administration and would normally seek to avoid matters of policy which are controversial between the major political parties. They would carry out valuable inquiries into matters of direct concern to many ordinary citizens, such as hospital administration, prison rules, training of teachers and agricultural research. Their reports would be fully argued and their evidence would be detailed, but we do not envisage that the deliberations of such committees would be reported or that they should debate publicly.

(c) Specialist Committees of Advice and Scrutiny should eventually cover the whole field of administration.

(d) five such committees might be considered as an initial experiment:

Scientific Development.

The Prevention and Punishment of Crime.

Machinery of National, Regional and Local Government and Administration.

Housing, Building and Land Use.

The Social Services.

(*e*) as new specialist committees are set up, however, the Estimates Committee should devolve its relevant functions to them in their respective spheres.

(*f*) the proposed specialist committees might eventually at least form the nucleus for the standing committees. We note other suggestions for committees to examine proposals for legislation before bills are drafted. Our specialist committees could perform such a function if desired.

(*i*) notwithstanding the above proposals for specialist committees, we are of the opinion that the House should make more use of the select committee procedure for *ad hoc* reports and investigation into matters of current concern, as was the practice in the last century.

(*j*) Ministers should not be members of specialist committees but could, of course, appear as witnesses before them.

7.3.5 A Select Committee on Civil Science

From the Parliamentary and Scientific Committee, Subcommittee on Parliament and Science, Memorandum Submitted to the Select Committee on Procedure, Appendix 3 to the Fourth Report from the Select Committee on Procedure, House of Commons Paper 303, 1964–5.

The Subcommittee had been set up in June 1964 to consider *inter alia*:

What can be done to improve the existing machinery to ensure that Parliament can establish more effective control over scientific and technological policy.

This memorandum, proposing a new Select Committee to consider only civil science 'undertaken with public funds and in support of public policy' appeared with the Study of Parliament Group's proposals [7.3.4]. Its wording reflects the thinking of the Subcommittee's chairman, Austen Albu, MP, who had earlier published a timely article in the journal *Minerva* [7.3.2] which caught the current mood of Parliamentary reform. It is noteworthy that when a Select Committee on Science and Technology was set up in 1966 its terms of reference and its subsequent inquiries proved to be much broader in scope than the Parliamentary and Scientific Committee were advocating.

3. Most of the proposals made to restore the power of Parliament as a means of criticising and controlling the actions of the Executive envisage radical innovations in procedure amounting to major constitutional changes. The Sub-committee decided to restrict itself to the means by which Parliament could be better informed on research and development in the civil field undertaken with public funds and in support of public policy.

4. The Committee recognised that science and technology are both means and objects of policy. Whilst it is not for the experts to determine policy, they can point out the probable effects of alternative choices, and they will have responsibility in connexion with the implementation of the policy adopted. It is for Government and Parliament to make the choice. If Parliament were to decide to take steps to improve their present system of Parliamentary control of policy and administration in these fields, for instance by means of a Select Committee, it would have to define its order of reference so as to try to avoid becoming involved in policy decisions.

5. In view of these considerations the Committee came to the conclusion that a start ought to be made by the setting up of a Select Committee to consider the reports of such bodies as the Research Councils, the Atomic Energy Authority and the National Research Development Corporation and also the activities of scientific research groups and establishments in Government departments, with a view to informing the House on their work and future. The Committee considered whether or not it should also, from time to time, make inquiries of an *ad hoc* nature, when legislation or some major administrative decision involving scientific advice was about to be introduced or was under discussion. The advantage of this would be that Members of Parliament would then hear something of the differing scientific and technical views which were matters of consideration in the Department concerned. This would make for more informed debate and differentiate more clearly technical from political issues. It was felt that a Select

Committee should have this power, but that it should be sparingly used. It was hoped that a practice would grow up of referring legislation involving scientific matters to the Committee, before or during its passage through Parliament.

6. Suggestions have been made, from time to time, that Select Committees should be furnished with expert staff. The experience of the Estimates and Nationalised Industries Committees has shown, however, that a body of Members of Parliament, served by Parliamentary Clerks who stay with a Committee long enough to learn its procedure, can make valuable reports on complex matters. Moreover there would always be difficulty in obtaining experts of equal calibre to the Government experts they were examining. This has been found to be the case in the United States, where the experts employed by Committees seem to take over the major part of an inquiry and to think it their duty to catch out the Government witnesses. The Committee therefore decided against expert staff, but agreed that experts should be called, when required, as witnesses. It is hoped that an expanded Library and Reference Service, with scientific staff, would also be available to assist the proposed Committee.

7. It is accordingly proposed that a Select Committee should be set up in the next Parliament with the following order of reference:

> That a Select Committee be appointed to examine the annual reports of the Councils of the Privy Council for Research, the Atomic Energy Authority, the National Research Development Corporation and other similar bodies, and also such scientific and technological matters as are under consideration in Government departments as a basis for legislation or the formulation of policy. That the Committee consist of thirteen members and that five be the quorum of the Committee. That the Committee have power to send for persons, papers and records.

7.3.6 Parliament and Science

From Nigel Calder, *New Scientist*, 28 May 1964, vol. 22, p. 535.

Calder's article was based upon a paper he delivered to the second Parliamentary and Scientific Conference, Vienna, 1964, organised by the OECD and the Council of Europe. In the passage preceding the extract Calder summarised the solutions then being offered to improve Parliamentary oversight over science and technology: more committees, more scientifically qualified MPs, better information services. While conceding that these are not without some merit, Calder advocated a much more radical approach by parliaments.

Excellent and necessary as these proposals are, I do not think they are radical enough. To suppose that parliament should deal with science on a largely non-partisan basis is to kill from the outset the hope that the consequences of science will receive the intense and wholehearted attention from politicians that they deserve. Of course, there may be common ground in respect for the freedom of action of the scientist and support for education, but the uses that we make of science ought to be controversial, for the following reasons:

(1) If politicians of different parties have genuinely different beliefs about the nature of the just society, they should be expected to favour or discourage different technical possibilities, or at least seek to implement them in different ways.
(2) There is often a genuine dilemma or uncertainty in technical decision-making, which affords a basis for legitimate debate.
(3) The decisions often entail a large number of problems and ramifications of either a technical or social nature, which can best be exposed by energetic advocacy and opposition.

In fact I do not think that parliaments should be looking wistfully for some little technical influence to mend their self-respect. They should instead be seizing on the greatly neglected business of mapping out a course for our scientific civilisation. It is a task for which parliaments, and behind them the political parties, are peculiarly suited, simply because they can bring together men and women with a wide range of knowledge and interests, who in turn can consult experts in any field. Indeed, I should like to think of a parliament as a kind of university dedicated to the study of complex social and technical systems.

How is this historic task to begin? I think that the natural place is in the policy-making committees of the political parties, with the blessing of the party leaders. They should set out to review some of the obvious trends in science and technology, through conservative, liberal or socialist eyes, and consider what tendencies they should favour and what oppose. Where that is not appropriate, they should ask whether their existing policies need modifying in the light of particular scientific developments.

I believe that, very soon, a successful start along these lines would generate a self-sustaining chain reaction of interest, in which parliamentarians, rank-and-file party workers, and sympathetic scientists and technologists would find themselves caught up. Because the universities and the bureaucracies have failed to put together a comprehensive picture of scientific change, which would provide a framework in which particular possibilities could be appraised, it may be relatively easy for parliamentarians to get ahead of the executive in this respect. By the nature of science, parliamentarians would have to be prepared to review their opinions continuously, and to avoid being dogmatic in areas where there may be great uncertainty. However, if a parliament showed itself to be possessed of a framework of ideas in the technical field, to be well informed on implications and to have a clearer sense of long-term purpose than the executive machine, I believe it would find a new power and enjoy new respect in the land.

7.3.7. The Select Committee System and the Power of Parliament

From House of Commons Debate, 14 December 1966. Speech by Rt. Hon. R. H. S. Crossman, MP, Hansard, vol. 738, col. 479–86.

Richard Crossman, Leader of the House of Commons, puts forward, in a debate on procedure, his proposals for an extended system of Select Committees to strengthen the powers of inquiry of Parliament.

The question the reformer has to ask himself is whether we should look backwards in an attempt to restore the pristine powers of this House to which our procedures are relevant or whether we should accept our present limited functions largely as they are and adapt our procedures to them. I know that there are some of my hon. Friends who dream of a time when the secret negotiations of the Government with outside interests which precede all modern legislation and the secret decisions in the Committee Room upstairs which largely determine party attitudes will be rendered insignificant because the House of Commons will once again become sovereign and make decisions for itself. I think they are crying for the moon.

It is no good trying to reform ourselves by harking back to ancient days. An effective reform must be an adaptation of obsolete procedures to modern conditions and to the functions we should fulfil in a modern highly industrialised community. Today, for example, it must be the electorate, not the Commons, who normally make and unmake Governments. It must be the Cabinet that runs the Executive and initiates and controls legislation, and it must be the party machines that manage most of our business, through the usual channels, as well as organising what was once a Congeries of independent back benchers into two disciplined political armies. Since this is the structure of modern political power, the task of the reformer is to adapt our institutions and procedures to make them efficient.

I believe that there are three questions by which the working of the House of Commons can be tested, both today and for future change: First, is the legislative process designed to enable policies to be translated into law at the speed required by the tempo of modern industrial change? Secondly, can our timetable, so long as the Finance Bill dominates our procedure, leave room for debating the great issues and especially for the topical debates on matters of current controversy which provide the main political education of a democracy? Thirdly, while accepting that legislation and administration must be firmly in the hands of the Government, does the House of Commons provide a continuous and detailed check on the work of the Executive and an effective defence of the individual against bureaucratic injustice and incompetence? It is by these three tests, I suggest, that we should try out both our existing procedures and the proposals for modifying them put forward by the various schools of parliamentary reform. . . .

It would be difficult to think of any democratic legislature where control of the Executive's expenditure by the legislature is less effective than it is here at Westminster. The British Parliament grew up with two great aims – to control Supply and to redress grievance. The machinery for the fulfilment of the first of these aims is now largely mumbo-jumbo and this despite the fact that with every decade the power of the State and the size of the public sector of expenditure grow steadily bigger. It seems almost true to say that the more of the taxpayers' money which the Government spend, the less effective is the check exerted by those elected to control it.

I have no doubt that in this respect at least the transfer of power from Parliament to the Executive has gone too far. We need to consider ways in which, while leaving the Executive the necessary freedom of action, we can develop institutions detailed, continuous and effective in their control. This reform would be good for the prestige of Parliament, good for the morale of back benchers, and, I believe, very good for the Government as well, because a strong and healthy Executive is all the stronger and healthier if it is stimulated by responsible investigation and criticism. That in a nutshell is the case for the two new specialist Committees which I am proposing to establish. . . .

I now turn to what is perhaps the most important aspect of parliamentary reform – for the reasons which I have given – the extension of our Committee system and, in particular, the extension of Specialist Committees along the lines of our successful Nationalised Industries Committee. I am aware that there is a body of opinion on this side of the House that our aims should be not to deflect the energy of the House into Committees upstairs, but to restore its vitality here on the floor.

Personally, I believe we can combine these objectives, by trying to get more and better controversy on the floor while reviving parliamentary control of the Executive through the Committee system. I have already alluded to the problems of the Finance Bill and I now pay tribute to the detailed post-mortem supervision now exerted by the Public Accounts Committee – which celebrated its 105th birthday this year - and to the Estimates Committee, too.

But should we go beyond the supervision of expenditure? Should we experiment in giving to the back-bench Member a share in the investigation of administration and even of the policies of the Government? I am convinced that this is a debt we must pay. Some people talk as though it were still possible to make a sharp distinction between policy and administration and to lay it down that Select Committees and specialist Committee should deal only with the latter and be excluded from consideration of the former.

But even a superficial study of the work of the Committee on

Nationalised Industries, for example, will reveal that this distinction between policy and administration is often very blurred, and that it is the blurring which enables the Committee to do its most valuable work.

There are countless areas of public discussion and public concern which are not areas of party controversy and it is these areas which Select Committees can profitably investigate. This will have great constitutional importance. For one thing, we have to face the growing power of the State, the ever-widening expansion of Whitehall's activities, the ever-enlarging number of civil servants and agents of the State. At present, huge tracts of this public sector are virtually screened from accountability to the House of Commons.

I believe that, cautiously and systematically, the House of Commons should begin to bridge this gap. That is why, after careful consultation with the Opposition, we shall follow up the proposal of my right hon. Friend the Prime Minister, in the debate on the Address, and establish experimentally, for this Session, two new Committees – one, a subject Committee on Science and Technology; the other, the first Committee to study a Department, the Department of Agriculture.

Before I leave the subject of Select Committees, there is one other point in the Motions which I shall be moving to which I call particular attention. It is that the two new Specialist Committees shall have power to admit strangers during the examination of witnesses unless they otherwise order. At present, as the House knows, Select Committees can sit in public but the right to do so is a little mythical, because the withdrawal of strangers may be enforced whenever a single Member spies them.

Under the arrangement which we propose, these two Committees will be able to hear evidence in public, unless at any time they prefer to sit in private; and I hope that this example will be followed in the Sessional Orders continuing the existence of our other Select Committees.

7.3.8 A Welcome for the Select Committees

From 'Chink in the Flood Gates', *Nature*, 24 December 1966, vol. 212, p. 1395.

In this leader the journal emphasises, first, the educative function

of the Select Committee on Science and Technology, suggests
some topics of immediate relevance, and looks forward to the day
when the Committee will seek to influence policy.

Mr Richard Crossman, the new Leader of the House of Commons,
has done much in the past week to win friends. He may yet influence people. The issue is the modernisation of the ancient and
archaic machinery of Parliament in the United Kingdom, and Mr
Crossman has done more than any of his predecessors in the recent
past to acknowledge that something must be done to improve on
the present state of affairs. It is true that he has not gone so far as to
suggest ways in which Members of Parliament might be given
decent facilities for doing a decent job, and it remains to be shown
that sittings of the House of Commons on two mornings each week
will prove more than an empty joke, but there is no doubt that his
proposals to establish two select committees on science and technology and on agriculture will be valuable precedents. In the long
run, they will provide the House of Commons – and also the other
interested parties, particularly the Cabinet and the Civil Service –
with an opportunity to learn by practical experience that detailed
inquiry by a subject committee can be a valuable means of
scrutinising what the Government is doing without being an intol-
erable interference with orderly administration. It is particularly
welcome that science and technology should be one of the first
subjects to be dealt with by the House of Commons in this way.

But what should the new committee do? What questions should
it ask? And how should it seek to influence policy? To begin with,
at any rate, it is to be hoped that members of the committee will
not attempt too much. If anybody thinks that the committee on
science and technology can be used as a means of forcing policies
on a reluctant government, he should be quickly disabused. And if
the analogy between the committee and the public watchdog is too
rigorously followed, the committee will frequently find itself frustrated by too much detail and too imperfect a sense of direction. In
this sense, comparisons with other legislative committees may be
misleading. The Public Accounts Committee of the House of
Commons has, for example, built up a splendid reputation for
counting candle-ends with precision, and in that incarnation has
become a fearsome institution throughout Whitehall, yet it is
rightly so much concerned with uncovering the procedures by

means of which past decisions were arrived at that it can only indirectly influence the forward objectives of public policy. Then the committees of the Senate and the House of Representatives in the United States most frequently function as statutory links in the machinery for approving the budgets of the several agencies of the Administration: they, too, spend a lot of time crying over spilt milk. The new committees in the House of Commons will be most valuable if they can seek to influence the formation of the principles on which new policies are constructed. But to begin with, at least, they will have to be satisfied with an educative role. In the immediate future, the most valuable task which the committee could attempt would be to bring out into the open the whole process of forming public policy in science and technology. This, after all, is how the Commons Committee on the Nationalised Industries has made its mark.

To this end, it is particularly welcome that Mr Crossman has it in mind that the proceedings of the committees should usually be open to the public. Although there are arguments the other way, only by such means can the committees hope to claim and to hold the attention of the outside world which the potential importance of their work deserves. The committee on science and technology should not, in practice, too frequently rush for the bolt-hole of secrecy which Mr Crossman has allowed for in his statement. It is also to be hoped that it will gather evidence widely, and from outside the Government and its immediate connexions. There is no reason why it should not frequently provide a forum in which critics of government policy might urge heterodox arguments – or even discredit themselves.

7.3.9 The Select Committee on Science and Technology: the First Round

From Roger Williams, *Public Administration*, Autumn 1968, vol. 46, pp. 309–11.

The 'first round' of the title refers to the completion of the Select Committee's first inquiry into the United Kingdom nuclear reactor programme (House of Commons Paper 381–XVII of 1966–7). In this passage, Williams attempts to assess the Committee's future role and examines the relevance of American experience with a similar committee concerned with atomic energy.

The hopes which the Minister of Technology and others have expressed about the Committee's future lead inevitably to the question of what role the Committee can best aspire to fill. Another Minister may well want to encourage the Committee at some future time to act as an outside assessor of a tricky sector of science or technology which he foresees is likely to require his initiative. There does not seem to be anything improper in this, provided that the Committee is not used as a mere sounding-board for a policy which the Minister has already decided upon, that it does not fall under departmental influence, and that it does not again have a subject forced upon it as was oil pollution. It must be in the much longer term when, if ever, the 'party nexus' has been replaced by a 'parliamentary nexus' that the Committee will be able to assume some legislative function.

Remembering the wistful envy which the Science Committee had for the American Joint Committee, it seems worth considering whether there are any respects in which the Science Committee could follow the latter's example. There are perhaps two features of the Joint Committee to be noted. In the first place, much of its prestige has stemmed from its low membership turnover compared with that of the US Atomic Energy Commission. Whether the Science Committee will have a relatively stable membership will presumably depend upon the satisfactions service on it is felt to provide, and also upon the number of MPs who are likely to be interested in its type of work. The satisfactions are as yet indeterminate, though the attendance of members during the first inquiry was high and there was only one change in the membership between the first and second investigations. What is certain is that the number of science-minded MPs is not, or not yet, large. The original membership of fourteen consisted of eight Labour members, five Conservative and one Liberal, and nearly all of these could claim some connexion with science or technology, either by education, by occupation or by previous membership of a party committee concerned with science. The Chairman, as well as having served for several years on the Nationalised Industries Committee, is also a chartered engineer and was in 1967 Chairman of the Parliamentary and Scientific Committee. If the Committee members elect to devote themselves to thorough examinations of important and often complicated subjects, and to build up their stock of knowledge by regular committee service, then the Committee seems bound to acquire the influence that accompanies experience.

The other feature of the Joint Committee's history which may be relevant to the Science Committee is that the Joint Committee has achieved most when it has been unanimous. In preparing its first report the Science Committee divided three times, once broadly on party lines. Whilst it would be undesirable and to some extent counter-productive (i.e. in relation to the Committee's function of stimulating discussion by going into all the arguments relevant to particular issues) for the Committee to feel compelled always to show a united front, there is little doubt that over a period of time recommendations expressing unanimous decisions are likely to carry most weight with the Government.

However, the immediate task for the Committee is surely to inform Parliament, and thereby the public, about the Government's role in science and technology and the impact of these on the broader affairs of government. The need for a wide knowledge of the processes and criteria of science and technology policy has already been recognised by political scientists and others. If the capacity of Parliament to help meet this need was ever in doubt, it ought no longer to be when the Science Committee's report had multiplied many times what was publicly known about civil nuclear policy and administration. Any select committee report which accomplishes this is invaluable. It follows that the quality of the evidence which the Science Committee elicits from its witnesses is of much greater significance than its own pronouncements on that evidence.

If this argument is correct, disagreements among members of the Committee are matters of concern only to the extent that they hinder the inquiry. As suggested above, a committee consensus would be an advantage in building up its influence. It might also relieve Parliament of the labour of working through the evidence itself. However, both of these desirable ends would depend upon the Committee's unanimity having been reached after a thorough probe as opposed to a superficial survey. In any case, the onus remains with Parliament to debate the reports of its select committees, whether or not these are unanimous. It is, of course, true that Parliament has frequently not used the reports for this purpose in the past – too often they have not been relevant to its main concerns, but it still seems reasonable to hope that as more and more information is prised from the departments by other specialised committees as well, then this must in time improve the quality and increase the diversity of parliamentary debate.

7.3.10 The Case for Preserving a Select Committee on Science

From House of Commons Debate, 29 October 1969. Speech by Mr Eric Lubbock, MP, Hansard, vol. 788, col. 1012–6.

The Commons were debating the report from the Select Committee on Procedure on the scrutiny of public expenditure and administration (House of Commons Paper 410 of 1968–9). This had recommended (para. 32) that the Estimates Committee should be changed to a Select Committee on Expenditure which should work through a series of 'functional' Subcommittees; eight of these were recommended, each to have nine members (para. 33):
Industry, Technology, Manpower, Employment
Power, Transport and Communications
Trade and Agriculture
Education, Science and the Arts
Housing, Health and Welfare
Law, Order and Public Safety
Defence
External Affairs.
Lubbock, the sole Liberal Member on the Select Committee on Science and Technology at that time, deplored the possible disappearance of that Committee and went on to suggest a still more radical change.

Mr Peter Emery (Honiton): Today's discussion is proving to be a debate in the best sense of the word, but there is one point which needs clarification. Paragraph 35 of the Report sets out the tasks of each of the subcommittees. However, in paragraph 34 one sees the proposed order of reference of each subcommittee. It says:

'To consider the activities of Departments of State concerned with [naming a functional field of administration] and the Estimates of their expenditure presented to this House; and to examine the efficiency with which they are administered.'

That seems to exclude——

Mr Speaker: Order. This intervention is suspiciously like a speech.

Mr Emery: It is not intended to be. That seems to exclude what has been debated across the Floor of the House – the specific item of policy with which the hon. Gentleman is as concerned as I am.

Mr Lubbock: I think that I have the hon. Gentleman's point. He is asking what is to happen if an item in which the subcommittee is interested does not appear in the Estimates. I can give the House several examples of that occurring in the context of the Committee on Science and Technology. If one refers to the Report of the Subcommittee on Carbon Fibres, there is a good deal in it about the policy on the licensing of know-how to companies in the United States. However, one will find nothing of that in the Estimates of the Ministry of Technology. Yet it was of vital national interest.

Another example is that of the subcommittee which has been reconsidering the nuclear reactor programme of the United Kingdom. That Report recommended that the Minister of Technology should consider guaranteeing the contingent liabilities of electricity boards which brought into operation new types of reactor systems. We could look in these forward expenditure projections to the next four or five years and, unless it was already Government policy to offer these guarantees to the electricity boards, nothing would appear. This was a suggestion which the Select Committee was making for the consideration of the Minister of Technology, but now he has to consider it in his dual capacity of Minister of Power as well.

If we started from the Estimates Committee we would never consider the matter in the first place. That is what I am trying to get across and what the hon. Member for Honiton (Mr Emery) has confirmed.

I come now to the second point, which is really very difficult for the hon. Gentleman to answer, namely, how a subcommittee of nine members will cover such an enormous sphere. I can speak only from my own knowledge, but the Select Committee on Science and Technology, which consists of 15 members, has such an enormous task that it has been found necessary to divide it into subcommittees for some of its investigations and still it has not been able to cover even a fraction of the area. Yet the hon. Gentleman proposes that a committee of nine members will deal not only with technology, but also with industry, manpower and employment. I do not see how it can possibly go into the subject in the detail that he suggests. I suppose it is feasible to divide a committee of nine members into two subcommittees of four and five members respectively. But then problems of political balance arise with the even numbers on the committee of four. Therefore, I think that it would tend to work in practice as a whole committee, as the Estimates Committee does. It does not divide into numbers smaller than six or seven, although the hon. Gentleman may be able to correct me on that.

The work of a committee of nine members covering the enormous spectrum of interests that the hon. Gentleman named will be superficial in the extreme, and as a result of its work there will be no control by this House over the activities of the Departments concerned. . . .

I come now to my third criticism. Specialist Committees have built up good relations with the outside world – at any rate, those that have been allowed to continue in existence for longer than one Session. If they are to become subcommittees of an Estimates Committee under a new name, no one will take them seriously.

I do not want to make a great thing about the prestige of the chairman of a Select Committee. Taking the Select Committee on Science and Technology as an example, the hon. Member for Bristol, Central (Mr Palmer) is heard with respect wherever he goes. People know that he has been doing this job since the inception of the Committee, and before that he was Chairman of the Parliamentary and Scientific Committee. The hon. Gentleman proposes that he should be called the Chairman of the Subcommittee on Technology, Manpower and Employment of the General Purposes Committee of the Expenditure Committee, or something of that kind. I do not think that the outside world will take such a position nearly as seriously or respectfully as the position of someone who is Chairman of the Select Committee on Science and Technology. The same applies to the other new specialist Committees which have been established with our full support.

I turn now to my final criticism of the hon. Gentleman's proposals. One of the most successful features of what I might call the post-Crossman Committees is their ability and determination to get stuck into policy – sometimes to the great embarrassment of Government Departments. We need only remember the first examination of the Agriculture Committee into the Common Market to appreciate that.

With shortages of manpower both in the Clerk's Department and among hon. Members willing to serve on these subcommittees, there will be a danger of creating so many new Committees if we go on with the specialist system that we embarked upon under the Crossman administration that we would not have enough people to go round and the committees would not do their jobs thoroughly. Therefore, I should like to propose a different solution. I suggest that we wind up the Estimates, Nationalised Industries and Public Accounts Committees and that we redistribute the resources to man as many non-Departmental functional Committees of the Crossman type as we can.

I agree that it is absurd to claim that the Estimates and Public Accounts Committees give this House any real control over expenditure. The Comptroller and Auditor-General could still make reports to the House on any items that he thought should be drawn to our attention. The House could then decide whether any of the allegations of the Comptroller and Auditor-General should be referred to the appropriate specialist subcommittees. For instance, advanced gas-cooled reactor royalties were considered by the Public Accounts Committee in its last report, as well as army boots and about 50 other matters. The report of the Comptroller and Auditor-General would, under my suggestion, be referred to the Select Committee on Science and Technology which would decide whether to consider it and to report on it to the House, or it might take note of it and leave the House to take such action as it thought fit.

It is absurd to pretend that the Select Committee on Nationalised Industries, which is to be kept in being under these proposals, can examine a range of industries from Cook's Travel to the British Steel Corporation. A vast spectrum of industries comes under its scrutiny now which would be far better dealt with by functional Committees, not of the kind that the hon. Gentleman wishes to create, which are components of a much larger Committee, but which are independent and as powerful in their own right as we have now.

We, on the Liberal bench, have always pressed for the extension of the Select Committee system to redress the balance of power between Parliament and the Executive. We agree that that should be the task of this House. That far we go with the hon. Gentleman, but we do not think that he has done it in the right way. If this

Report is accepted, the Executive will have no difficulty in keeping the watch puppies absolutely docile and we will be creating a new piece of House of Commons bureaucracy which deals mainly with the Treasury – and we know how good that is at avoiding awkward questions.